THE INVENTOR'S GUIDE
How to protect and profit from your idea

About The British Library

The Science Reference Information Service (SRIS) is a world leader for information in science, technology, medicine, business, patents and the social sciences. As part of the British Library, SRIS provides a unique resource for the UK science, business, government and academic communities by offering access to a wide range of literature, including the largest patent collection anywhere in the world.

Specialist information services in business, science and technology, the environment, health care, patents and social policy are able to give expert help to users. They offer free, quick enquiry services for simple queries and competitively priced research services for more complex enquiries.

To complement the services of SRIS, documents can be supplied to customers offsite via the Library's Patent Express service in London (for UK and overseas patents) and via the Document Supply Centre (DSC) in Yorkshire. DSC provides the most comprehensive photocopy/loan service in the world from its unrivalled collections of journals, books, conference proceedings and theses.

For more information on the Library's collections and specialised services, please fax SRIS (44)(0)171 412 7947 or DSC (44)(0)1937 546333.

THE INVENTOR'S GUIDE
How to protect and profit from your idea

David Newton, Roy Fuscone, Michael N Russell, Martin Jones, Jeremy Phillips, Charles Dawes, Richard Paine, Sheena E Will and Nigel Spencer

Edited by The British Library Patents Information Team

THE BRITISH LIBRARY

Gower

Published by Gower Publishing Ltd
Gower House
Croft Road
Aldershot
Hampshire GU11 3HR
England

Gower
Old Post Road
Brookfield
Vermont 05036
USA

The editor and the contributors have asserted their right under the Copyright, Designs and Patents Act 1988 to be identified as the editor and authors of this work.

British Library Cataloguing-in-Publication Data
A catalogue record for this book is available from the British Library.

ISBN 0 566 07994 1

Typeset in Bembo by Concerto and printed and bound in Great Britain by MPG Books Ltd, Bodmin, Cornwall

Contents

Foreword

Dr John Ashworth, Chairman of the British Library

To achieve success every inventor requires some luck – in practically every invention serendipity plays a part – but more importantly the inventor requires advice and information. This book aims to put the reader in touch with sources of both sound advice and good information.

To bring an invention to fruition requires input of knowledge in many different fields, from science and technology through intellectual property law, business planning and marketing to product design and production. Few inventors possess this breadth of learning themselves or, indeed, have easy access to all the many areas of expertise required. Some will have access within their own organisations but for those that do not, the United Kingdom is well endowed with institutions, organisations and individuals who are able to provide assistance to the diligent inventor.

The British Library is an institution that can and does help inventors. It is the world's leading resource for scholarship, research and innovation and sits at the hub of the national library provision in the UK. It has unrivalled collections of scientific, technical and business literature and offers services to help the inventor find and use this information. The later sections of this guide explain how to use the Library directly or through local public and university libraries.

This book cannot guarantee inventors success, whether in the form of hoped-for wealth, recognition, or the satisfaction that comes from making a useful contribution to society. The advice of the authors and the contacts given should, however, enable you to determine, sooner rather than later, the future prospects for your own invention.

Armed with the knowledge of where to seek out help and information you will still need exceptional fortune as well as perseverance and hard work. So... Good Luck!

How to Use this Book

This guide is intended to cover many of the topics of use to an inventor in taking, or planning to take, an invention from the initial germ of an idea to the final product or process. Inevitably some topics are covered in less detail than the reader may require but the guide also has, as a key feature in Chapters 7 and 8 and in the Appendices, details of libraries and organisations where further help may be obtained, lists of other sources and a bibliography.

This book has been written by a number of different authors, all expert practitioners in their particular field, and each chapter may be read independently of the others. There is some overlap in the coverage of the chapters to make them self-contained and aspects such as confidential disclosure, for example, are mentioned in several places. In addition, to help the inventor who wishes to dip into the book for particular subjects, the text includes cross references to related sections of other chapters.

For the inventor who wishes to read through from beginning to end to obtain an overview of the processes necessary to reach commercialisation the book has been arranged in a logical sequence for planning purposes, if not necessarily for action. First, is there a market for the invention? (Chapter 2); second, how can its development be financed? (Chapter 3); then, how can the idea be protected from exploitation by others through patent and other intellectual property rights? (Chapters 4 and 5); and lastly, how is development, prototyping and manufacture best carried out? (Chapter 6). For an inventor wishing to take an invention through to commercialisation for the first time, attention to all areas is necessary and a read of the entire book is recommended.

At the end of each of these chapters the reader will find a section containing key organisations able to provide help related to the subject of the chapter. The full contact details of these and other organisations are given in Appendix 1, Organisations. The Appendices also give details of the books mentioned in the text along with a number of other useful sources. The institutions, companies and other bodies listed in this book are given as examples of sources of more detailed help and information, and details of additional bodies offering help in the same fields can be found in the publications listed in the bibliography or in other directories. The inclusion of names in this book should not be taken to indicate an endorsement of the services offered.

Chapters 7 and 8 are designed to provide a signpost to sources of information, both technical and commercial. They describe sources and list libraries throughout the UK and online services available anywhere from which key information may be obtained.

The book was as up to date as possible at the time of final editing in 1997 but please bear in mind that some information may have changed by the time you read this book.

Notes on Contributors

Chapter 1: David Newton

David Newton heads the Patents Information services at the British Library's Science Reference and Information Service.

Chapter 2: Roy Fuscone

Roy Fuscone consults in the management and exploitation of intellectual property including the establishment of new businesses. He is accredited by the Chartered Institute of Marketing, an Associate Member of the Chartered Institute of Patent Agents, and a Fellow of the Institute of Patentees and Inventors. He is a director of Marketing Aids (UK) Ltd.

Chapter 3: Michael N Russell and Martin Jones

Michael N Russell and Martin Jones are both directors of TIE (UK) Ltd, a company which is largely concerned with technology transfer on a worldwide basis. Michael Russell has many years' experience in raising capital for innovation both from within a venture capital company as well as establishing and co-ordinating the European Venture Capital Associations and the European Commission's seed capital network. Martin Jones, who founded TIE (UK) 12 years ago, has assisted many innovative companies from Europe and North America in promoting their know-how by means of strategic alliances, licensing, or other routes.

Chapters 4 and 5: Jeremy Phillips

Jeremy Phillips writes and lectures internationally in many fields of intellectual property. He is currently intellectual property consultant to London-based solicitors Slaughter and May, editor of the *European trade mark reports*, the *Butterworths intellectual property law handbook* and the *Aslib guide to copyright*, as well as consultant editor of *Managing Intellectual Property*.

Chapter 6: Charles Dawes and Richard Paine

Charles Dawes and Richard Paine, who have many years' experience in the successful international launch of innovations and new product ideas, are directors of Inventorlink. They are backed by the full services of the Inventorlink research, marketing, PR and licensing departments which together have achieved many licensing/royalty agreements and also outright sales, joint venture agreements and new company formations.

Chapter 7: Information officers at the British Library

The staff who have contributed to this chapter are: Sue Ashpitel, Armen Khachikian, Ian McKevitt, Steven van Dulken and Sheena E Will. They are all information officers working for the Science Reference and Information Service, where, with their colleagues, they answer thousands of enquiries every year with the help of the information sources detailed in these chapters and in the Appendices.

Chapter 8: Nigel Spencer

Nigel Spencer spent seven years as an information officer in the

British Library Business Information Service and now manages the British Library Science Reference and Information Service Holborn Reading Room.

1
Introduction – the Importance of Information

David Newton

"To reinvent the wheel" is a commonplace expression and one that is easily comprehended. Reinvention is all too frequent an occurrence but one which can and should be avoided by the inventor. Not only in the invention process is it possible to repeat unknowingly what has already been done by someone else but also in the chain of processes that the inventor has to follow to put the product or service successfully into the marketplace. While it may be entirely desirable to repeat what someone has done before, in some spheres it is essential, or at least highly desirable, to know what others have done before so as to be able to follow or to improve upon best practice.

Some inventors may suggest that to investigate what others have done will stifle creativity – but to know what has gone before creates a base level upon which to improve. How can an inventor think laterally or creatively without knowing what the current knowledge base is and what the current problems in need of solution are? In any case most innovations are small improvements on what is already known and so require a sound knowledge of the present state of the art before that element of creativity can be added by the inventor to produce the improved product or process.

Scientific data will often be meat and drink to the inventor. Few would think of re-measuring common scientific and engineering coefficients, other physical properties or chemical properties, but much other data has already been determined by scientists and engineers in the past. This data is often timeless although it may be re-determined with greater accuracy as measuring techniques and instrumentation improve.

Aside from setting a base-line for improvement, some information is essential for survival – in more senses than one. No one would deny that it is necessary to employ safe practices and follow the relevant laws in the workshop or laboratory, relying where appropriate on the experience of others. However, for the continued existence of the invention as a product or business it is also necessary to be aware of those laws which govern commerce and ways of doing business.

Not all of the information required for the successful commercialisation of inventions is published or even recorded on paper, but a huge and growing amount is. This book gives help on where to find that published information whether it is in print or in an electronic format.

Throughout the process of innovation, knowledge of a wide range of human endeavours is always desirable and sometimes absolutely necessary. Information alone, though, does not always prove sufficient to enable the problem to be solved, because understanding is also necessary. On occasion the hours spent trying to master a subject will still be insufficient to give understanding of a subject and at this point it will be necessary to seek expert advice. Advice is essentially information adapted to the current circumstances by an expert, but professional advice can be costly. Background information on the context can usually help to determine what the adviser should and should not be asked to do and so improve the communications between the parties. It is not that a little knowledge is a dangerous thing but the inventor does need to recognise the extent of the knowledge base and his or her own grasp of it.

Since the quality of the advice provided by a professional adviser cannot be checked directly when it lies in a field of expertise unfamiliar to the requester, it has to be tested indirectly by, for example, checking the professional qualifications and standing of the adviser, receiving a recommendation from a third party, requesting information on work carried out previously for other clients, or asking the right questions from a better knowledge base as suggested in the previous paragraph.

The quality of information obtained from elsewhere can be variable and, as with advice from professional experts, it can be critically important to the success of the invention. Checking the worth of printed information is often difficult and perhaps for this reason it is often left undone. It is frequently assumed that the quality of any textual source is unimpeachable, but this is clearly not always the case. The strength of an article in a learned scientific journal is that it has been refereed by an expert in the subject and papers in the journal should be of high quality as a result of this process. A small proportion of journal articles are subsequently reviewed and these reviews may provide a useful analysis or criticism of the article. Alternatively a study of subsequent citations of the article, e.g. using *Science Citation Index*, can reveal comment on it.

Books may perhaps be judged by the status of the author or the author's institution and usually there is more substance on which to corroborate any assessment. With abstracting periodicals and databases the original source is usually readily available as a check. Information contained in patent specifications should be of a high quality as the applicant will normally wish to ensure strong patent protection through meeting the legal requirements for an adequate description. However, there is nothing to stop an applicant setting out false information if he or she is willing to pay for it and is not concerned to receive a valid patent. Nor is the fact that a patent has been duly examined and granted any guarantee that the information contained in it is sound. Finally, the quality of information gleaned electronically from the sources such as the Internet will be very variable, perhaps more akin to conversational chatter in some instances.

Ultimately the suitability of information for its intended purpose rests with the user because quality embraces accuracy, currency and relevance. Only the user can judge the relevance and the importance of currency. In this book the advisers, information suppliers and other sources are listed without an implied guarantee of their quality, completeness or soundness.

It has to be recognised that there is a cost involved in searching out information even though, in technical fields, this is rarely of the same order of magnitude as undertaking research directly. There are many systems which help improve access to information and perhaps reduce this cost, at least in terms of time if not in money, starting with the public library itself as an institution which collects and organises material, through library classification schemes and catalogues to computer indexes and databases. But information is sufficiently important that plans need to be made to ensure resources are allocated to this part of the invention process. This book seeks to enhance access to the information and, through improved knowledge of where to look and what to expect, to enable plans to be made for its acquisition. Ideally an information audit should be conducted to determine what is already known and what still needs to be acquired.

Technology continues to change the way information is disseminated. The volume of information available via the "Information Superhighway", the Internet network of computers, is doubling in size every eight to ten months according to some reports. In the US the White House is pushing vigorously for its development, which it claims can help unleash an information revolution that will change forever the way people live, work, and interact with each other. In the UK this network is already at least as accessible as anywhere else in the world.

While electronic sources of information are widely recommended throughout this book, the library is still usually the best place to start gathering information (and often the place for all information needs). The function of the library is to bring together particular groups of sources and this, combined with the catalogues and the expertise of librarians, is still a very powerful system for accessing information. Increasingly, as the value of information is recognised, some library and information services are seeking to charge for their services. While the impecunious inventor may not wish to use charged service, others may find, as with the use of other forms of advice, that the purchase of services can be cost-effective and speed the process of getting the information required.

This book is a result of the strongly held belief that information is vitally important for the inventor. It is hoped that, for those who share this belief, this text may provide help in locating, accessing and understanding the ever increasing pool of available knowledge.

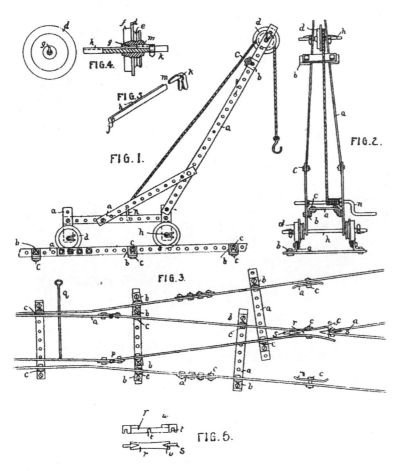

FIG.4.

FIG.5.

FIG.I.

FIG.2.

FIG.3.

FIG.6.

*Figure 1. British patent number 587/1901. Frank Hornby's
Meccano patent, applied for in 1901.*

2
Marketing an Invention

Roy Fuscone

This chapter aims to help the inventor answer the crucial question, "Can I make money from my invention?" It should also enable the reader to plan exploitation of the invention through licensing, sale, or by setting up in business.

The chapter is divided into five parts:

- preparing the ground, which shows the need for confidentiality and asks some basic questions about whether an idea has good potential
- detailed evaluation, which describes how the technical and commercial viability of an idea may be assessed and lists the main factors that determine success or failure
- towards exploitation, which describes the alternative routes to exploitation, discusses the involvement of marketing consultants, and explains the preparation of marketing documents
- technology transfer, which deals with exploiting an invention through a technology transfer contract
- starting a business, which covers the main factors involved in setting up one's own business in order to exploit an invention.

2.1 Preparing the ground

Patent protection is important, complex, and costly, so it is essential to arm yourself with a good understanding of what is involved: Chapter 4 discusses this in detail.

2.1.1   Disclosure of the invention

It is generally unwise to tell people about your idea as soon as you have it. Before telling anybody, you should take into account the likely consequences. For example, you can only obtain a patent for an invention if, among other things, the invention is new at the date the application is made. By disclosing an invention to someone, other than in confidence (see below), you may disqualify yourself from obtaining a valid patent on an application filed after that disclosure.

Patent protection

The filing of patent applications must be planned carefully. If you file too early in the invention's life, the patent could run out at an important stage during the commercial exploitation of the invention, leaving it unprotected. Patenting internationally must also be undertaken as part of a detailed plan.

Patent agents and solicitors to whom you disclose your idea will treat it in strict confidence; they are bound by a professional code of conduct and such a disclosure will not affect its patenting. You should always seek confirmation from other advisers that they too agree to accept your disclosure in confidence. (A sample confidentiality agreement is given in Appendix 5.)

Many companies refuse to accept an unsolicited idea in confidence for various reasons which are aimed at protecting their own interests, not yours. If you are thinking of approaching a company with your idea, the less detail disclosed about its technology the better. It should always first be established under what conditions the company is prepared to receive your information. You or your exploitation agent (see Section 2.3.1) should negotiate a confidentiality agreement with the company to cover the disclosure you wish to make. The same caution should be applied to anybody else who is not bound to keep your secret.

Confidentiality undertakings should cover documents, drawings, models, prototypes and samples, together with any other format in which confidential information is contained. A solicitor should be consulted by you or your agent before you enter into any disclosure agreement, to advise as to its effectiveness and as to its commercial desirability.

Finally, even if you are confident necessary confidentiality safeguards are in place, check any existing intellectual property rights before disclosing your idea. Are they properly identified, documented and recorded in case you have to defend them against infringement?

2.1.2   Ownership

Although you have had an idea, you may not be its legal owner and may not therefore be entitled to exploit it independently, or at all. If you are an employee of a business, your position depends upon your contract of employment, and you should familiarise yourself with Sections 39–43 of the Patents Act 1977. If you are employed by a university your position may depend upon the statutes of your institution as well as the contractual terms of your engagement. The industrial liaison officer of that institution may be able to advise.

2.1.3   The inventive concept

To get the best out of your idea it is important to identify the essence of your invention. Ask yourself these three basic questions:

- what does the idea do?
- what is important?
- what can be omitted?

Suppose for instance that your idea concerns a fisherman's folding stool and that you have either sketched it in a drawing or made a working model. The essence of the invention could be in the way the collapsible frame, the seat, or the legs are constructed, or how they work together. The advantage of the finished product could be a stool which is lighter, more compact, or stronger, or a combination of these.

Assume that it is in fact the way the frame folds and unfolds which makes it special (i.e. that this is what the invention does); you next need to decide which parts of the frame are essential, and which are not, for producing this operational effect. When you have identified them, go on to ask:

- are there alternative ways of putting these essential parts together?
- are there alternative forms for these essential parts that could still work to produce the same operational effect?

With the answers to these questions, you have a much clearer understanding of the inventive concept which is contained in the drawing or working model. More significantly you have separated the inventive concept from the actual product that originally came into your mind. This is something you should always try to do if at all possible. Only when you have decided upon the inventive concept can you begin to develop one or more product concepts.

2.1.4   Applications of the concept

Normally one inventive concept has several different applications or fields of use. You should produce a list of all the possible candidates for yours by answering the question: in what other products could the inventive concept be incorporated?

Taking the example of the stool, other products such as a table, a workbench, an ironing board, a different type of chair, a bed or indeed any product having a collapsible structure could, in principle, incorporate the inventive concept that has been identified.

2.1.5   Viability

Now that you have defined one or more inventive and product concepts, you can move on to the next stage of examining the commercial viability and the technical viability of each of the product concepts on your list.

It is best to carry out this work in ever increasing detail with a GO or STOP decision being taken at appropriate points. This means that you will only go on to the detailed work described in the next part of this chapter, if you are satisfied that the results of the initial work justify the investment in time and probably money.

The following five headings, three of which relate to the marketplace, illustrate the types of questions that usually have to be considered at this stage. Investigating the marketplace for your product is crucial, for, ultimately, no matter how ingenious your idea, it is only there that money will be earned. Only by assessing markets will you be able to identify openings, if any, for marketing your product commercially, whether by first or third party exploitation.

Market demand

- who are the end users for this product?
- is the market international, national, regional or just local?
- how big is the market for this sort of product?
- will the number of end users in this market increase or decrease over the next five to ten years?
- how long do products last in this market, both in terms of buying frequency and the period between one product generation and the next?
- is the market subject to changes in fashion or craze? (What is in one year may be out the next in the case of children's toys or gimmick products.)

Market resistance

- how familiar are potential end users with this sort of product?
- are the advantages and benefits of the product immediately obvious to an end user?
- would the product require the end user to change or learn new skills to use it?
- does the potential end user need the product?
- would the end user have to use the product with another product?

Market conditions

- are there any laws and standards that the product would have to satisfy before it could be sold?
- how safe is the product for the end user or consumer?
- could the product harm the environment?
- would the product have to be repaired by the end user?

•   would spare or replacement parts have to be supplied?

Commercial viability
•   how practical is the intended function of the product?
•   can the product readily be manufactured?
•   how much time, money and other resource is needed to develop and test the product?
•   how much would it cost to make the product?
•   how much would it cost to distribute and sell the product efficiently?
•   how likely is the product to be profitable?

Market competition
•   why and how is the product better than competing or substitute products?
•   which firms compete in the market?
•   at what prices are they selling their products?
•   should the product be protected under the intellectual property system? If so, how?
•   would the product infringe somebody else's intellectual property rights?

2.1.6   Is protection available for your idea?

If you have protected your idea by keeping it secret or confidential, and your basic research indicates that it may be worthwhile to protect it and try to market it, you will need to explore the intellectual property rights system. This is a complex legal system which protects intellectual property by granting statutory rights of various sorts to the owners of the property. Depending on the circumstances in which the invention or other intellectual property came into existence, the owner may or may not be the inventor, designer or creator. Both the originator and the owner – if separate – may be entitled to benefit from the successful exploitation of the property.

There are a number of different types of intellectual property right, of which the most important are:

- Patents
- Trade marks
- Design rights
- Copyright.

A number of these may be relevant to your idea. They are discussed in more detail in Chapters 4 and 5.

Before you decide to commit time and money to trying to protect your idea by use of the intellectual property rights system it is wise to learn as much as you can about it. As well as reading Chapters 4 and 5, ask the Patent Office for their free information pack. Chapter 7 gives advice on how to check whether your idea seems novel. If, after doing this, you want to go ahead and try for patent protection, you should seek the advice of a patent agent, and, where appropriate, a trade mark agent. The section at the end of Chapter 4 tells you how to get help with this.

Should you be dealing with an idea for which you may wish to register a design you can get advice on your chances of success before committing yourself. In this case the course of action currently recommended is to apply for the Design Registry at the Patent Office to do an infringement search against your idea. You should apply on the appropriate official form with the necessary fee, providing a drawing of suitable quality for use if you subsequently wish to apply to register your design, or a photograph of the artefact you have designed. If you are advised that your design would not infringe existing rights it could be worthwhile to proceed with an application.

2.2 Detailed evaluation

The next section (2.3) helps you with the decision on whether to start a new business or license your invention to an existing company, but you are not yet ready for that step. There is no point in trying to prepare a plan of action until you have a clear understanding of how an idea evolves into a product ready for launching onto the market or the factors that will largely determine its success or failure following market launch. This section discusses this, and the difficult hurdles your idea must clear if it is to earn money.

2.2.1 The success rate

The statistical probability of a raw idea eventually turning into a money spinner is small. Preliminary evaluation of their viability shows that most ideas, whatever their technical excellence, are not worth investing in. Technical and commercial viability are what count in the question "is this idea worth taking seriously?"

Some products which are initially promising fail later on. This may be due to an insoluble problem cropping up during technical testing and development work. Of the few that do mature into new products launched onto the market, it has been estimated that between 30 per cent and 50 per cent fail in the sense that they do not produce a reasonable return on the investment made in them.

These hard facts of commercial life are only too well known to companies, banks and other sources of capital and make them extremely cautious in deciding whether to back an idea for either first or third party exploitation.

Studies of why companies experience such a relatively high failure rate with new products indicate that the main factors at play have nothing to do with the technical ingenuity of the originating idea. Rather they stem from the circumstances in which the idea was taken up and exploited, as can be seen from the following list of factors contributing to failures:

- inadequate market research and analysis to identify the real needs of the intended customers and the strength of the competition
- technical deficiencies in the product, usually as seen through the eyes of the user or the consumer
- lack of effective marketing effort when the product was launched
- higher costs than expected
- reaction of competitors and their competitive strength in the marketplace
- poor timing of product launch
- technical or production problems.

The main factors which led to the success of other new products have been found to be:

- the products were market driven – in other words they were designed to meet a previously identified and clearly defined end market
- the customer's needs, wants or preferences were satisfied by the products
- there was effective and efficient management during the innovation phase (i.e. the development of an invention into a product)
- considerable marketing effort was made to launch the product.

2.2.2 New product management

Businesses succeed with new products largely because, apart from managing them efficiently, they:

- create a pool of new ideas
- make a preliminary assessment of their viability
- select those few that appear potentially profitable
- define the selected idea in terms of a new product concept
- define the market gap which the new product is intended to fill
- agree provisional specifications covering the new product, its market, how it is to be manufactured, and its marketing and selling
- test and develop extensively – this can include trial selling
- finalise the specifications referred to above
- manufacture efficiently
- market and sell professionally.

They also know that, in average terms, if it costs £1 to create the raw idea (invention), it can cost £5-£10 to turn it into the final pre-production model and £10-£100 to set up the production, marketing and sales facilities to make and sell the product commercially. Though these are general average ratios for a hypothetical product rather than realistic sums, they show just where the balance of the investment will lie in exploiting your idea.

Companies also often engage teams of professionals, including outside consultants, in evaluating an idea and defining the provisional and final pre-production specifications. Team members will include technologists, production engineers, designers, salesmen and accountants.

Obviously an inventor is most unlikely to have all this expertise. Nevertheless you should realise that your idea must meet the same exacting standards and pass through the same procedures if it is ever to earn money for you. It is essential that you undertake the best possible preliminary assessment of your idea before you do anything else toward its full exploitation.

2.2.3   Preliminary assessment

Normally the preliminary assessment of an innovative project is undertaken in several stages, of increasing detail, at the end of each of which a GO or STOP decision will be taken. The criteria used to make assessments differ according to the nature of the assessor's business and its future strategy. A large company might set higher minimum limits for revenue streams from new products than would a smaller company, but may be content with lower margins of contribution to overheads and profit. An assessment may be conducted in the following stages.

Stage 1 Compatibility
- does the proposed new product fit within the company's strategic business plan?
- does the new product's market fit into the markets to be addressed by the company's plan?
- is the new technology needed to develop the product within the company's planned scope?
- is the production process needed to make the product within the company's planned development?

If you are thinking of promoting your idea to a company, note how much importance is given to an idea having to fit in with the company's plans for the future. No matter how good your idea is, if it doesn't suit the company, it is most unlikely to be of real interest.

Stage 2 Feasibility

- are the design and development of the product technically feasible?
- has the company the resources to undertake the technical development involved in producing the final product?
- is it technically feasible to make the product?
- are the resources and facilities available to make the product (or can they be made available)?
- has the company the resources and organisation to sell into the target market?
- could the product easily be copied by competitors?
- is the product capable of development into generations of product or of differentiation into a range of products?

Answering these questions will inevitably be an imprecise science and it is just not possible to "spot the winner" every time. Some "good" ideas will always slip through the net but this leakage is commercially acceptable for any business which has a large pool of raw ideas to draw upon.

Stage 3 Attractiveness and opportunity cost

If the project passes these first checks, further criteria will be applied to assess its "attractiveness" along these lines:

- what are the product's advantages and benefits to the end user or consumer? (e.g. in terms of price, quality, function and novelty)
- what are the advantages to the company of the product's market? (e.g. in terms of the market's current size, future growth, strength or weakness of the competition, and the ease of addressing and penetrating it)
- how does the particular project compare with other potential developments in the company's activities?

- what is the opportunity cost in following this particular project through to commercial maturity, i.e. what else could the company achieve by applying elsewhere the resources needed for the project?
- would successful completion of the project result in a long or short term advantage to the company and would this be consistent nationally and internationally?

Stage 4 Detailed business plan

The project will go to a third evaluation stage, with greater emphasis on financial planning to encompass projections of likely investments, expenses, revenues and profits. (The business plan is discussed in Section 2.5.1.) Only if it then gets the "go" will work begin on its commercial exploitation by the company.

All these stages are centred on the basic question: "Is the idea viable?" You must be convinced that the answer is yes – and you must be able to demonstrate the probability that this is correct to anyone whose support you seek – whether you opt to back the project yourself or sell or license it to another party. Otherwise your time would be better spent in devising another idea. There are many sources of help with drawing up a business plan: one of the many excellent handbooks for those starting up in business, free advice booklets from the large banks, or personal help from the business information centres described in Chapter 8.

2.3   Towards exploitation

If, having taken stock of your idea and applied professional attitudes to assessing its viability you have concluded that it is worth your while trying to exploit it, you must decide which of the three following routes you are going to take:

- license it under a written contract which is commonly called a technology transfer agreement (third party exploitation by licensing)
- sell it outright to a third party (third party exploitation by assignment)

- keep it to yourself and exploit it through your own enterprise (first party exploitation).

Before you can make a decision to go along any of these routes you may need to raise funds to enable you to develop drawings, models or prototypes, and to arrange and conduct trials. In addition you may need funds to secure and maintain any relevant intellectual property rights.

If your idea is well researched and presented, and you have arranged for appropriately skilled people to work with you, you may be able to obtain funding for your basic endeavours. All aspects of finance for the inventor are discussed in Chapter 3.

2.3.1   Exploitation agents

You will need to consider whether to engage the services of a reputable exploitation agent to act on your behalf, or to exploit your idea yourself. An agent may seek an initial fee or time fees as well as a share of any money you make from an exploitation that the agent arranges. (This topic is also discussed in the next chapter, in the context of funding the development of your invention.)

The main advantages of using an agent are that:

- having experience of the evaluation of new products, the agent should be able to assess yours better than you can
- the agent may have client companies seeking inward licences and will already know what they are looking for (or would consider)
- the agent will know how to present an idea to a company in a professional and credible manner
- the inventor is free to get on with other inventions or business.

If you are thinking of engaging such an agent, satisfy yourself that the agent is competent, reliable and is offering you a service that meets your needs and interests. It is possible to make enquiries at the Chartered Institute of Marketing, which keeps a register of officially accredited consultants in the many aspects of marketing.

You can also seek guidance from the government-supported business information services now established throughout the country, described in more detail in Section 8.7.7. There are also a number of organisations offering a range of help to inventors seeking to develop their ideas: the Institute of Patentees and Inventors, and the Intellectual Property Development Confederation, are two of these.

Some agents, as well as trying to match inventors with companies, also offer services to help them raise venture capital and most of them will undertake some initial assessment of an idea as part of their service. Even if you use an agent, it is always best to carry out your own preliminary assessment of the idea, to find out whether the idea justifies paying for professional advice.

The results of your assessment should help an agent provide a more cost effective service to you if you do engage one.

2.3.2   Choosing a champion

Even if you are developing your invention yourself, it is frequently essential to find a "champion" – someone who is prepared to provide the backing you need to get the idea launched commercially. This champion is likely to be the representative of either a company with which you wish to collaborate under a technology transfer agreement or a bank or some other source of capital from which you want to raise finance.

In either case you will have to choose the most likely candidate for your champion and to prepare marketing documentation that will whet an appetite for a meeting at which your idea can be presented, together with a proposal for its exploitation.

2.3.3   Identifying potential exploiters

Work done to assess the idea will help you and/or your agent to decide which companies should be approached. In researching the marketability of the idea, you will have discovered which companies are already serving the target market or could do so. These will often be the first targets. Ideally you are looking for a company which:

• has a good track record in exploiting inventions and will deal with them under confidential disclosure
• is of a size that matches the scope of your idea in terms of how much it is going to cost to exploit and its profitability potential
• has a business structure which will accommodate your idea
• has clear business objectives but is not necessarily the market leader
• is not fully committed by dint of its current activities and can thus apply time, money and manpower to your idea.

Your professional advisers such as your patent agent, solicitor or accountant may be able to help, since one of their other clients might be a firm looking for new ideas. There will almost certainly be some centre for promoting business enterprise in your area. If there is a Business Link (in England), a Business Shop (in Scotland), a Business Connect office (in Wales), or a Local Enterprise Development Unit (LEDU) office (in Northern Ireland), this is the place to put you in touch with all the local business opportunities. More details are given in Section 8.7.7. Elsewhere there will be some of the following: a Chamber of Commerce, Training and Enterprise Council, Local Enterprise Agency, a public library with business information, a Job Centre.

There are also a number of regional offices of European-wide initiatives to encourage innovation. Innovation Relay Centres (IRCs) and Business Innovation Centres (BICs) have slightly different remits. The IRCs were set up primarily to encourage activity in established businesses and research teams, but their interest in technology transfer could provide an opening for an individual inventor with a suitable product. The BICs offer a range of services to help both businesses and individual inventors. There is also some European support for projects run by other organisations offering help to inventors, such as the DEVICE programme at NIMTECH. This assists inventors to get their ideas into the marketplace quickly and with maximum effect, covering all aspects of product development, from technical problems and patent protection to market assessment and finding commercial backers. Information about the major regional networks such as NIMTECH and WEMTECH is available from the business information centres mentioned above.

2.3.4   Marketing documents

The big advantage of preparing marketing documents is that it helps you clarify your own thinking on the proposal you want your champion to take up. In addition it is businesslike to provide an opportunity to consider your proposals before meetings are held. It is essential to make a good first impression, and to elicit a serious, useful response to your plans you should start by introducing your project profile.

The profile
The purpose of the profile is to bridge the credibility gap that you will almost certainly encounter when seeking the backing of your selected champion. A champion is likely to start by assuming that your idea is not to be taken seriously, particularly if you are a first time inventor. The document simply has to make him or her want to meet you and explore your proposals further.

The profile should be:

• typed
• clear
• well written
• concise (no more than two pages and preferably only one)
• businesslike.

Do not simply send a copy of your patent application or other detailed documentation under cover of an introductory letter. Your profile should briefly summarise:

• the technological and commercial environment which give rise to the opportunity for a new product
• the inventive concept of your idea in the broadest of terms

- the products to which this concept could advantageously be applied
- the markets for these products
- the main features of these markets
- your proposals for the product i.e. licensing or a joint venture
- why the person reading it should be interested in the proposal.

Clearly the contents of this profile will be based on the results of preliminary assessment of the idea; these contents will not be expressed in detail but in an appealing, highly general summary. In preparing it you should assume that the reader will not know anything about its subject. It has to be both simple and self-explanatory.

In dealing with the reasons why it should be taken seriously, you should also draw attention to a more detailed document – a presentation – that contains further, particular details of the new product proposal summarised in the profile, and of the work done to establish its viability.

The presentation

The fully documented presentation of your idea, and its development and testing up to the point where you believe it constitutes a viable commercial proposition for an exploiter, should set out the following:

- the availability (if any) of a prototype, diagrams, engineering drawings, photographs and anything else which shows forms of the idea
- reports concerning the feasibility of the idea, e.g. market survey reports or the results of technical and/or user tests that you have made or commissioned
- the names and addresses of the professional advisers and product testers you have used to date
- the results of searches to check on the novelty of the idea, the scope for its protection and on whether its exploitation would infringe somebody else's intellectual property rights (if no such clearance search on this has been made, an indication should be made that it is still outstanding)

- the intellectual property rights you control in relation to the idea to date, as well as the present possibilities for adding to these in the future (e.g. the filing of patent applications overseas using the priority date arrangements in the Paris Convention)
- your inventive concept and product concepts
- the markets in which these concepts are to be exploited
- financial projections showing assumptions for:
  - addressable markets in units of "product"
  - market penetration percentage(s) to be achieved
  - product unit direct costs
  - product unit ex-works price
  - likely product unit end user price
  - product unit royalty sum and/or percentage of ex-works price
  - manufacturer's investment in manufacture
  - manufacturer's investment in marketing
  - gross revenue stream
  - royalty stream
  - contribution to manufacturer's overheads and profits.

On the basis of such assumptions negotiations may be conducted to reflect the actual business plan for the new product, which the licensee is invited to adopt and which an eventual licence must reflect. Such modelling exercises can also be used to establish the value to a manufacturer of an assignment under outright purchase rather than a royalty licence. An exploitation agent may be able to provide advice on these procedures.

The fully documented presentation should provide annexes of information from reliable sources to support and lend credibility to all assumptions and interpretations given in the main argument.

Rather than send your detailed presentation document to potentially interested people, you should seek to arrange a personal presentation of it by you or your exploitation agent. At such a presentation your models, prototypes and other supporting material should be tabled.

The making of commercially useful presentations is a professional activity that an exploitation agent may be able to perform more effectively than you could. If you or your agent cannot arrange a presentation, you will have to rely on submitting your marketing documents. Prepare a suitable covering letter explaining briefly who you are, enclose your profile and presentation documents, and suggest a meeting at a mutually convenient later date. The letter should also state the address and telephone number through which you or your agent can be contacted.

In any document describing your idea and any work done to establish its viability, and any covering letter, you should avoid exaggerations, expressions of belief or unsupported opinions, jargon or obviously biased views.

Finally, the documentation should be carefully reviewed in case improvements can be made, before it is utilised. The profile document will be non-disclosing in nature but your presentation document may disclose proprietary information. Such information, whether contained in documents, drawings, samples, models, prototypes or other materials should not be disclosed unless a confidential relationship has been established.

2.4 Technology transfer – licences and assignments

Any legal right that you have in an idea can be sold or shared. If you sell the right completely, the agreement under which ownership of the right is transferred to the buyer is called an assignment. If instead you only give your permission to somebody to use the right while retaining ownership, the agreement is called a licence. (Licences are also discussed briefly in the context of financing innovation in Chapter 3, and in the context of patent protection in Chapter 4.)

A legal right can come from owning any of the following kinds of intellectual property:

* a patent, for an inventive concept in a product or process
* a registered design for the eye-appealing features of shape, configuration, pattern or ornamentation of the product

- a registered trade mark for selling the product, i.e. a brand name or symbol
- a copyright and an unregistered design right in an artistic work concerning the product, e.g. engineering drawings illustrating it and its component parts
- a copyright in a literary work concerning the product, e.g. a computer programme for making it, a production or workshop manual
- know-how, e.g. a piece of secret technical or commercial information relating to the design or manufacture of a product.

Applications for intellectual property rights can also be transferred, as can the right to make these applications. In many cases, one transaction will cover several types of intellectual property (e.g. one or more patents, a trade mark, copyrights and confidential information).

Because all these rights stem from a country's national law, they must be owned under its law before they can be transferred, and such transfer will be limited by the country's national frontier. So if you are thinking of exploiting your idea internationally through one or several arrangements, you will need to decide:

- in which countries the idea is worth exploiting
- in which countries you want to pursue first party exploitation and in which third party exploitation
- in which countries you can afford to apply for intellectual property rights in order to build an appropriate intellectual property portfolio
- in which countries you could allow your licensee(s) or assignee(s) to acquire further intellectual property rights over your idea such as patents.

The central point is that to structure a marketing campaign, you must first decide upon plans for the exploitation of the idea and the acquisition of the intellectual property rights you need.

The filing of foreign applications for patents and registered designs, as well as the communication overseas of unpublished technological information, may be subject to various legal controls. Your patent agent will be able to advise you about these.

2.4.1 Advantages and disadvantages of licensing

If you are not already in business, are not interested in setting up to exploit your idea commercially and do not wish to sell your idea outright then licensing is the only route to take. This is not limited to the UK; indeed it is wise to investigate which markets provide the biggest opportunity for your idea internationally, and the costs and benefits of addressing these.

If you already run a business but are unable to take your idea any further forward because of lack of time, money or other resources, licensing should be considered. In this way you may still be able to make some money out of your idea. Take for example these situations:

- you want to exploit the idea internationally, but lack the means for doing so. Instead of trying to set up an exporting system, perhaps with overseas agents, one or more licences with local foreign firms might be appropriate.
- if you are thinking of exporting to countries where there are exchange controls, import duties or other restrictions, licensing local firms might be the best way over these obstacles. The same is true if freight costs would make exports from the UK uncompetitive.
- a joint venture with another firm may provide risk sharing
- it may be better for technical reasons to collaborate with another firm in the joint development and testing of the idea; if not the manufacture and sale of the end product
- it may be possible to obtain another firm's technology by way of a cross licensing deal; you obtain the right to use their technology in exchange for their right to use yours.

In summary, licensing makes good business sense in many instances. It can:

- reduce the cost and risks to you of exploiting your idea alone

- allow you to exploit markets that would otherwise not be addressable
- help you earn money from your idea more easily and quickly than if you started from scratch.

Particularly for companies, it may offer the prospect of additional advantages such as:

- higher sales of spare parts, raw materials (if the idea contains a process), semi-finished products, and sometimes even production plant and machinery
- entry into new markets, both domestic and foreign
- a sharing of risks, technical knowledge and experience regarding the performance of the licensed product
- the exchange of valuable commercial information on changes in local market conditions, customer requirements or new business opportunities.

The main disadvantage with licensing is that you share your idea and any profits it earns. Therefore, if a promising idea fits in with its overall business strategy, a company will usually prefer not to license it to another company.

Further disadvantages with licences stem from the fact that they are a sort of contract. There is always the chance of the other side not keeping to the agreement, or of circumstances changing so drastically that the agreement becomes a poor deal for you.

If you are a private inventor with just a raw invention, the main problem with licensing is finding and persuading a company to come to an agreement with you. With no other bargaining counters, you may be asking the company to take all the risks but only a share of the rewards. To obtain such a deal, not only must your idea be seen as a likely winner, you also have to market it and yourself very well indeed.

2.4.2   Advantages and disadvantages of an assignment

The principal advantage of concluding an agreement to assign your rights in your idea to someone else is that you will receive your payment when the contract is signed, or in agreed stages. This may suit you much better than waiting for payments of royalties over time. In addition you will probably remain free of any further expense in relation to your idea. Under the terms of a licence agreement you may be responsible for the maintenance of intellectual property rights, not to mention the costs relating to any disputes in connection with them.

Although you may receive less in payment for an assignment than the potential value of future royalties (20 per cent to 50 per cent of the cumulative royalty stream is not unusual), you will have funds available for early use. The main disadvantage of an assignment is that products based upon your idea may become very successful. Royalties that would have been payable to you under a licence may turn out to be much more lucrative than an assignment value which would normally be based upon only agreed minimum royalties.

You must choose between a bird in the hand and perhaps two in the bush in the light of your personal plans and objectives.

2.4.3   Negotiating a licence or an assignment

The main factors to be taken into account when negotiating the terms of a technology transfer agreement are discussed below. It is essential to consider them before you prepare a proposal for submission to a company with which you wish to contract.

Any company that shows an interest in your idea will ask what sort of deal you have in mind. Your profile documentation should have made this clear. Unless you have already formulated your initial bargaining position, and how to justify it, you will not establish credibility for your proposals. The chances of achieving a deal could well be ruined.

The golden rule that applies to all contracts is that they should be long in the negotiation and short in the signing. There is no substitute for thorough and careful planning before one sits down at the negotiating table. Even so, there is no guarantee that negotiations will conclude with a deal. There is certainly no such thing as a standard contract; they are the products of hard and lengthy bargaining.

To safeguard your own interests, to increase the chances of the negotiations going smoothly and to make sure that you end up with a satisfactory arrangement, you may need to engage the services of a reputable exploitation agent and a competent solicitor. There is no point in making a deal only to be told by your advisers that it is impossible to formalise it with a binding contract, or that such a contract would be illegal under European competition law rules. By being involved from the start your advisers will be in the best possible position to negotiate a contract for your signature.

2.4.4   Key components of a licence

The following paragraphs deal only in outline with key components of a licence; a satisfactory licence will inevitably contain a number of further provisions.

Who are the parties to the agreement?
*   you must be sure that a licence is signed by properly authorised representatives of a company and that this authority extends to all and any type of organisation with whom the company is directly connected such as holding companies, subsidiaries, associated companies.

What intellectual property rights do you wish to license?
*   you may wish to specify patent applications, copyright etc.

Is the licence to be exclusive or not?
*   you could agree to a simple (or non-exclusive), sole or exclusive licence. If you wish to reserve the right to exploit the licensed rights yourself and to license them to further licensees, then you are dealing with a simple licence. If you do not wish to license them to anybody else but also to exploit them yourself, it is a sole licence. If you want to agree to license them to one licensee but to no other and not to exploit them yourself, it is an exclusive licence. These three types of licence give the licensee increasing degrees of security against competition over the exploitation of the rights, which in turn affect your ability to secure a higher price.

What is the territory to be covered by the licence?
- in terms of manufacturing, and/or selling and the development or exploitation of future products.

Under what conditions must the licensee exploit the licensed rights within the licensed territory?
You may want for example:

- to set minimum quality and/or quantity conditions which the licensee must meet when making and selling the product
- to impose conditions on the licensee for keeping any licensed confidential information secret
- to license the licensee to make just one of the products covered by the licensed patent application or other intellectual property rights
- the licensee to use your name in selling or advertising of the product
- to stipulate that the licensee must not grant sub-licences.

On what basis are payments to be made to you by the licensee ?
You may want:

- an "up-front" payment when the licence is either signed or becomes operative, as well as further payments afterwards
- fixed sums to be paid at regular intervals during the life of the licence
- payments to be made on the basis of a percentage of the ex-works selling price of the licensed product.

How are you going to make sure that the licensee performs its obligations?
- you must ensure that the licensee does its best to make the most of the licence and so earn you the most money.

You may want:

- to impose "best endeavours" commitments
- to stipulate minimum sales quantities for royalty accounting.

How are improvements to be dealt with?
You may want:

*   to have a share in any inventions which the licensee creates from yours
*   to include or exclude any improvements you may make.

Who is going to pay for any legal actions against alleged infringers of the licensed intellectual property rights?
*   are the rights to be insured against infringement and if so at whose expense?

Who is going to maintain the licensed intellectual property rights?
You may want:

*   the licensee to pay the costs of obtaining the grant of intellectual property rights or of renewals
*   the licensee to pay to obtain intellectual property rights in countries covered by the licence.

What happens if the licensed rights expire or are invalidated?
*   your licensed applications may fail the examinations required by the local law under which they were made
*   the validity of the licensed rights may be challenged successfully by a competitor; who will pay to defend them?

When and under what provisions does the licence come to an end?
You may want:

*   the licence to end before the maximum possible life span of one or other of the licensed intellectual property rights
*   the licence to end if the licensee fails to keep the licensed confidential information secret
*   the documents containing the information returned to you with a promise that the licensee will stop using it when a licence comes to an end.

What happens if a dispute breaks out over the licence?
You may want:

- the licence to specify that any dispute is to be settled through arbitration
- if it is international, the licence should specify the national law of a country under which it is made.

2.4.5   Assignments

If you decide to sell your idea outright under an assignment, the contract to be negotiated will be somewhat different. It will still be necessary to ensure that the fundamental requirements for the existence of a contract are met, i.e. the parties must have the capacity to contract, there must be consideration in money or money's worth, offer and acceptance thereof must be explicit and performance of the terms of the contract must be implicit.

2.4.6   Give and take

The paragraphs above show the complexities of arranging a licence or an assignment. If a contract is made, it will certainly be a compromise compared to any original opening gambit, especially if you have nothing more to offer than a raw invention and a pending patent application.

You should always be prepared to accept that some cake is better than none. Do not make the mistake of refusing to consider and accept anything less than what you want. Every deal requires give and take on both sides, just like any good partnership; the other side of the negotiating table will be watching closely for this essential flexibility.

2.4.7   Option agreement before licence or assignment

It is not uncommon to find that even if your marketing documents and/or personal presentations have made sufficient impact to interest a target licensee in your idea, an option to license or purchase may be required as an interim step. Just as you should have a clear idea of your intended terms for licence or assignment, so you should have possible option terms prepared.

An option constitutes a contractual agreement and must therefore fulfil the requirements outlined above. It will normally allow the potential manufacturer to research and test your idea and to examine the validity of any intellectual property rights you may claim. In addition it must be made clear what happens as and when the option is exercised or allowed to lapse. If, for example, the exercising of an option is intended to lead to the signature of a licence, then the terms of that licence must be agreed between the intended parties when the option is signed. Indeed the option should refer to the intended licence and have a copy of it attached.

If your idea is sufficiently attractive to potential licensees and dependent upon the point you may have reached with intellectual property right applications, you may find it advantageous to grant multiple concurrent options. An exclusive or sole option could command a rather higher price than a simple option.

2.5   Setting up in business

The golden rule about going it alone is that you should not. If you wish to set up in business and need to attract financial backing you will need the advice and support of people with specialised training.

You can begin by seeking a free counselling session with a counsellor from any of your local advice centres. What is available varies from area to area. In many parts of England Business Links bring together all the most important support services in a single location, and offer the services of a Personal Business Adviser. Where this is not available the nearest Chamber of Commerce or Training and Enterprise Council is a good starting point. Section 8.7.7 has more information on the network of over 200 Business Links in England, and the similar Business Shops in Scotland, Business Connect offices in Wales, and LEDU offices in Northern Ireland. Counsellors from these organisations are usually experienced businessmen who can provide you with valuable advice and assistance. They will also be able to tell you about the numerous schemes for helping small businesses on their way, and your local Job Centre or Training and Enterprise Council will have details of courses available to you.

The bibliography in Appendix 8 gives details of some of the many books with helpful information for anyone thinking of starting in business, and it is worth checking the selection at your local central library. More specialist libraries and information centres are discussed in Chapter 8.

When establishing a new business organisation you must become acquainted with the statutory requirements that will apply to its administration and to your conduct of its affairs in your position of responsibility as a director or partner. Although professional advisers will be involved in the formal processes of your business management (e.g. your accountant in the preparation of your annual accounts) it is you, the principal in the business, who remains responsible and accountable under the law for the proper conduct of your business.

By dint of your own expertise or that of an exploitation agent and/or marketing consultant you will have prepared an initial document to outline your idea and your proposals for its exploitation.

If you decide to exploit your idea yourself, the profile will again serve as your introductory "feeler" to be put to potential sources of finance, and to be kept readily to hand by your professional advisers as an aide to discussions they may have on your behalf. It will also need to be supplemented by your business plan.

2.5.1   The business plan

A business plan is essential if capital is to be raised for any project. There are various potential sources of investment or loan finance (banks, venture capital houses, wealthy individuals). There are also a number of sources of financial assistance in grants, awards and so on, so it is necessary for your presentation to make it quite clear what you are offering to those who may invest in you or your idea. The business plan must give the following information on you, your background and your planned activity:

- the kind of organisation you want to establish – a sole trader, a partnership, a co-operative or a limited company are the main alternatives
- premises
- production machinery and equipment
- suppliers
- finance required and how it would be spent
- a detailed programme for:
  - development of your intellectual property rights
  - product development
  - product and market testing
  - production
  - marketing
  - sales and sales service
  - a timetable for implementing the business plan.

Sample outlines and checklists of what a business plan should cover are available from some banks and in many of the guides to setting up in business. In preparing the plan you should not be over optimistic; allow for things going wrong and indicate how these contingencies would be handled if they arose. Use as a basis the results of the preliminary assessment or stock taking of the idea that you have already undertaken. The more detailed and thorough this assessment, the easier it will be to prepare your business plan.

The most important subject of all concerns people. There are always many good new business propositions on offer to sources of finance. Many of them come from established businesses wishing to expand, diversify, make acquisitions or in some other way alter their structure. What established businesses have, that you as a newcomer do not, is people with relevant experience of the commercial world.

You will be competing for credibility before you can even discuss your financial needs. If you are to have any hope of attracting sound and adequate finance you will have to demonstrate that your business will be adequately staffed and managed.

Depending upon the nature of your new business and what

skills you yourself can bring to bear, you will need to show how the core activities will be controlled (marketing, sales, finance, administration, legal matters, and how the management team will be co-ordinated so as to plan ahead).

If you intend to start out with a small internal team supported by external experts and then to recruit as your business grows, your plans must reflect your anticipation of these changes. This means that you must be committed to diluting your personal share of ownership in the interests of growth. Inventors often fail to realise this and base their thinking on a desire to own and control a business completely. Above all, be realistic!

2.6 Professional organisations

- The **Chartered Institute of Marketing (CIM)** is the largest professional body for marketing. The CIM consultancy service covers corporate strategy, marketing and communications, either on a direct project managed basis or by matching clients to approved consultants selected from its own register.

- The **Institute of International Licensing Practitioners (IILP)** is a body founded to maintain professional standards of consultancy in licensing and related technology transfer activities. For a small fee the Institute will circulate details of products or technology for which licensing is sought to their members, who then contact the enquirer directly.

- The **Licensing Executives Society Britain and Ireland (LES)** is a body which brings together people with diverse interests in this area of technology transfer. Their secretariat welcome enquiries about the Society's activities, and have a list of members who can give advice on a professional basis.

- The **Association of British Chambers of Commerce** will provide enquirers with the address of their local Chamber of Commerce, which will have information and advice about business in their area.

- The **Patent Office** has produced an interactive self-help laser disc package to guide inventors through the whole process from the initial idea to exploitation. The package is called *Making It Happen*, and is available in a few Business Links and universities.

• Further information on specialised business information and services can be found in Chapter 8, which includes a discussion of the **Internet** as a source of business information in Section 8.7.8.

Contact details of all the organisations mentioned in this chapter are given in Appendix 1.

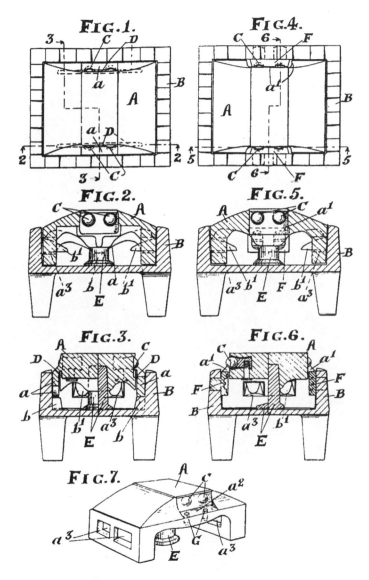

Figure 2. British patent number 457536. Shaw's Catseye road
reflector studs patent, applied for in 1935.

3
Finance for Developing and Marketing an Invention

Michael N Russell and Martin Jones

This chapter deals with funding all aspects of invention, and covers the differing circumstances of inventors already possessing some financial backing, and those with none. The chapter is divided into the following sections:

- introduction
- finance for the invention
- what is to be funded?
- sources of funding
- raising the funds
- managing your investors
- the future.

3.1   Introduction

Why does the inventor need more money?

A facetious question that appears hardly worth posing but, in fact, does require a serious and carefully considered answer. Before even beginning to attempt to answer it the inventor must consider the situation of the invention as a brain child that has been born but now needs food, shelter and clothing to sustain it, all of which have to be paid for.

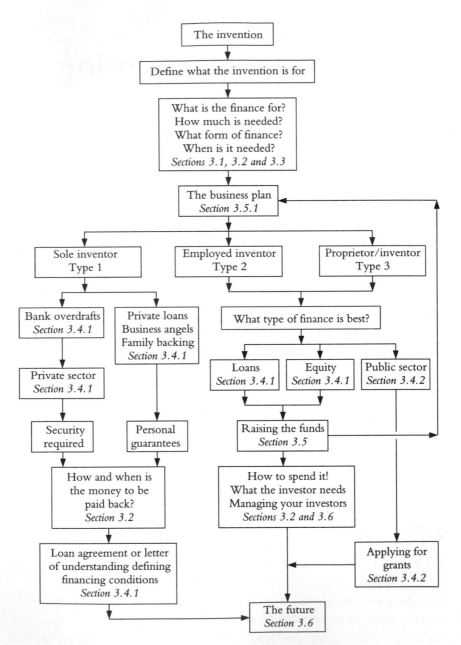

Figure 3. Routes to securing finance.

We will assume for the purposes of this chapter that the inventor comes in one of several guises, since the appropriate nature of finance and the way it is to be obtained will depend to a very great extent on the inventor's situation. We will consider three broad and general categories, namely:

- the sole inventor, with or without dependants to support
- the inventor as an employee, who will use the invention as a springboard to independence
- the inventor as the owner or founding member of an existing company who will use this company as the vehicle to launch the invention.

Why is it necessary to establish these categories? Figure 3 shows that the nature and type of finance that can be raised and the routes to securing this finance may differ considerably according to the present and future business status of the inventor and it is very important that the inventor avoids trying to raise the type of funds that are incompatible with his or her circumstances. A mismatch of funding type results only in disappointment and a waste of the inventor's valuable time; in extreme cases, it could prove to be the fatal injection which kills the invention. So always remember that it is vital to match your invention with the type and amount of funds sought. Nevertheless, much of what we have to say about this will apply to all three categories; we will draw attention to any differences where they become of overriding importance.

Another word of caution: whilst inventors will naturally be well experienced in science or technology they may have a much more limited experience of the intricacies of finance and the language spoken by financial wizards. Consequently on their first attempt to raise finance, inventors must realise that they are at the bottom of the learning curve. We hope that in this chapter we can move you quickly and effortlessly along this learning curve. Always remember, the ultimate success of your invention will depend on getting the right financing, in the right amounts at the right time.

3.2   Finance for the invention

It is most important to understand how much it is going to cost to develop the product or process to a marketable state and to set out from the start to obtain sufficient funds. Thinking of the invention is only the first step and probably the cheapest! Having got the idea, the inventor must first test its feasibility, prepare the patent application and carry out trials on the device before entering the production engineering and marketing phases.

All these steps costs money; the amount depends on the project and also on the route to commercialisation (see next section) that you, the entrepreneur, decide to take. Unless you have in-depth knowledge of the potential market for the invention, a market survey may be necessary as well. This could be a cheap and cheerful look at the potential for the product or process, or a sophisticated, detailed examination of the proposed market and the level of competition in it.

A prototype to test the feasibility of the device might be built relatively cheaply but it has to be sufficiently robust to stand up to testing. Without such testing it will be difficult to attract funding for the further development of the project. And for field trials of the product or process, it may be necessary to make several versions of the prototype.

Before any public display of the invention, either in trials or for fundraising purposes, it is wise to consider the possibility of patenting it to safeguard your rights. A basic preliminary application at the UK Patent Office is fairly cheap, but preparing the final text and extending the application to other countries will be very expensive. Appendix 4 gives an idea of typical basic patenting costs in the UK. You need to consider whether the market for the invention is sufficient to produce an income which will justify these costs. The writing of patent applications is a skilled art and inventors are strongly advised to take professional advice. The next two chapters deal with the legal aspects of protecting intellectual property, and the section at the end of Chapter 4 gives details of professional organisations to contact for further information.

The most expensive stage of all is likely to be the production engineering stage and this is almost always under-estimated by entrepreneurs. As a rule of thumb, moving from the prototype to the production model will cost two to three times the cost of getting to the prototype. There is advice on this stage in Chapter 6.

Finally, the cost of marketing the product must be taken into account. It is impossible to say how much each of the stages will cost as this varies greatly with the nature of the product. A better mousetrap or a simple mechanical or electronic improvement on current practice might cost only a few thousand pounds – perhaps only hundreds – to get to the production stage and could be developed in weeks. A highly complex process such as xerography took 25 years and millions of pounds to develop to the commercial state.

Capital for the invention can be provided in a variety of forms:

• equity capital by means of the sale of shares in the invention whether organised privately or through venture capital (we consider this later in the chapter)
• grants which involve an outright payment against certain targets or milestones
• loans of various kinds either privately provided or from banks or other financial institutions
• seed capital, a variant of the equity capital approach, providing money for the pre-start-up phase, often from family and friends.

Any new development with an inventive step will take many months if not several years' effort to bring to market and before it begins to make money for the inventor. Therefore the application of any outside source of finance that has to be sought to develop an invention must be very carefully timed. The capital will have to cover expenses throughout the development period, and the estimate of the likely rate of progress should be realistic, if not downright pessimistic.

Every inventor is faced with the problem of needing more time than was expected to bring an invention to a satisfactory conclusion. Every delay that occurs brings in its train a demand for more money! Making the capital stretch over an expanding work programme and an ever-increasing length of time really does need the skill of a tightrope walker.

When negotiating virtually every kind of financial package to provide capital the following key dates and actions need to be kept in mind:

- drawdown of the financing (staged provision of funds at particular milestones)
- interest payments
- capital repayments
- final discharge of the loan
- loan conditions.

These will all need to be defined in the loan agreement drawn up between the lender and the inventor before any funds can be released.

Commenting on these points in turn:

3.2.1   Drawdown of the financing

Unless the sums concerned are small, the lender will probably agree to provide the money in stages. Ensure maximum flexibility in the agreed availability dates and never bring into the project any capital until it is really needed. Unused capital is money in the bank for a rainy day! Avoid being too severely constrained on this point, and do not allow the lender to force on you any money that is not yet required.

3.2.2   Interest payments

Always pay the interest due on or before the required dates, never be late with interest payments. The inventor who delays such interest payments immediately destroys any confidence that the lender may have in the project. Before signing the loan agreement make sure there is a very clear statement on how any changes in interest rates will affect your payments and also on the method being used to calculate the interest you will be paying. At some time during the loan you or your accountant may need to check that the interest charges are being correctly applied.

3.2.3   Capital repayments

Do try your best to keep the capital repayments on schedule. In special circumstances the lender may agree to postpone such re-payments by three or six months to cater for a specific unforeseen difficulty.

3.2.4   Final discharge of the loan

Before signing the loan be absolutely clear about the conditions to be applied on discharging the loan. Sometimes an early discharge is advisable if your invention has taken off earlier than expected, so it is better to avoid penalties for a loan repaid ahead of time.

3.2.5   Loan conditions

Finally, be clear on the conditions under which the lender could call in your loan. Remember it may be advisable to have insurance to cover this if it includes *force majeure* clauses covering events outside your control.

When negotiating the loan conditions keep in mind the advantages of a longer repayment period, say five years rather than two or three years. Over such periods of time circumstances in technology and the market may change in unexpected ways which lead to a longer development phase.

It is also advisable with a longer term loan to define at least one loan review meeting with the lender. As mentioned above this gives you the chance, if circumstances change, to explain how you now wish to re-apply the financing. If you are well ahead or in advance of your projected progress the lender may even feel confident enough to extend the credit he or she is willing to offer you.

3.3   What is to be funded?

At this point, the most important decision for you is how to proceed.

There is no hard and fast rule to say which route is the best; it depends on your own inclinations and on the invention itself. The strategic choices range from licensing the rights in the invention for a fee and a royalty on sales, to setting up a complete manufacturing plant with all the attendant costs and complications.

We shall start first with the most difficult cases and then consider some financially less onerous alternatives.

3.3.1   Manufacturing

Setting up a business for the development, manufacture and marketing of the product or process is by far the most difficult option and likely to be the most difficult to get funding for. Lone inventors should not normally consider this; it is a route which should only be undertaken if there is a management team in place with sound knowledge of all facets of the manufacturing and marketing processes.

3.3.2   Assembly

An alternative to complete manufacturing is a "screwdriver" plant, where sub-contractors are used to manufacture parts which your factory assembles into a complete product, tested and marketed. This is in effect a low-risk route towards manufacturing, which will require less funding in its early stages. It still requires a management team but the costs will generally be much less.

3.3.3   Licensing

Licensing the rights to an invention is one of the most effective ways of obtaining an income from it, but will still require access to a certain level of funding. It is often the most appropriate route for lone inventors because there will be no requirement to set up a company or a team.

This is also an excellent route to commercialisation for academics and for industrial companies with intellectual property that is not being used.

All intellectual property – patents, trade marks, copyright and so on – carries certain rights to the exploitation of the invention and these rights can be traded like any other property. (The next two chapters discuss intellectual property rights in more detail.)

Inventors are in a strong position if they have a good patent to a desirable invention, but other forms of intellectual property, including even know-how, can be also licensed. The services of a good licensing consultant who can advise on a suitable strategy will help to avoid the pitfalls which lie in wait for the unwary, especially where know-how licensing is concerned.

The services of consultants do not come cheap but may prevent expensive mistakes and the total cost will in any case be far less than would be required to develop the invention by any of the routes outlined earlier (but please refer to Section 5.3).

Alternatively, you might consider setting up a licensing company to exploit your talents. This is probably most appropriate where there may be an "invention stream" or where the interest lies primarily in the creative process rather than in the manufacture and marketing. A dedicated R&D company may design items for clients ("contract R&D") or may develop them specifically to be exploited by licensing.

The factors affecting the decision to take this route are beyond the scope of this chapter, depending as they do entirely on your creativity and marketing skills. However, the costs involved in setting up such an operation are likely to be large; negotiating licences takes time and the income stream takes even longer to develop. You therefore need substantial working capital to sustain the company until the income stream builds up. (Licensing is also dealt with in Section 2.4 which discusses in detail the pros and cons of the licensing and assignment of intellectual property.)

3.4 Sources of funding

Funding can be obtained from a variety of sources, both private and public and you should first consider what is best for you: loan or overdraft, equity investment or grant. Not all of these will necessarily be available to you, of course.

3.4.1 The private sector

Funding from the private sector falls into different categories, each appropriate to the particular project in hand. In order of increasing cash requirements, they are:

- family and friends
- banks
- venture capital
  - conventional venture capital management companies
  - the informal sector (business angels).

It is worth defining a few terms at this point.

Seed capital
Money required to develop the invention, prepare a business plan, do market research and so on, before a company is formed. This is the seed stage and money going to the inventor will usually be in the form of a loan, secured or unsecured. The loan may be repayable at a later stage, or convertible into equity when a company is formed.

Start-up capital
Required to set up the company, hire staff, buy capital equipment and provide working capital.

Early-stage investment
Most companies require additional money after start up and this is especially true if the company is becoming successful. A company at this early stage is probably not yet profitable and still very vulnerable.

Looking now more closely at the various categories of investor:

Family and friends
Inventors looking for help to develop their ideas often approach family and friends for the finance they need. Normally such finance, if available, will be cheap, but the problem is that such sources are rarely able to provide more than the basic seed capital required to get started. Valuable though this may be it is unlikely to be sufficient to get the project into production and into the market. But if you do choose this route, remember that, even with family, you should prepare a brief document (a simple loan agreement) that ensures that all the parties understand the conditions under which the money is being made available, and which is signed by the lender(s) and the inventor.

Banks
The first port of call for increased financing for the start-up phase is usually your bank. Banks can provide either business loans or overdraft facilities, but while such loans may be appropriate for the short term, they are likely to be expensive. A number of banks do operate schemes to help entrepreneurs and it is worth enquiring about these. If it is not your own bank, you may be asked to give them your new company's banking business. Remember, too, that even with such schemes, the bank is almost always going to require collateral and you should consider carefully before risking the loss of your house or other security.

Venture capital
For larger sums, whether for start up, or for further financing, you should consider equity investment. Venture capital has developed a long way in recent years and is now an important source of equity capital. The advantage of this kind of investment, also known as risk capital, is that it imposes no charge on your company in its early vulnerable period. The venture capital manager will consider your business plan and assess the potential value of your company and then make an offer to buy shares in the firm at a price based on his or her valuation. No venture fund wants to create subsidiaries, so it is very rare for them to look for a controlling interest; more usually, the holding required is 20 per

cent to 35 per cent depending upon their view of the risk involved. The great advantage to you is that they will provide the money you need without security and, because it is not a loan, usually without interest as well. Their return comes with their eventual divestment, usually three to five years down the line, in the form of a capital gain.

Many people are reluctant to part with equity in their companies, and of course it does mean that you will have to share the profits in due course, but the best venture managers will be partners in your business and offer added value in several forms: financial or marketing expertise, international contacts and so on. Studies have shown that the hands-on assistance provided can make the difference between success or failure of the enterprise and it is better to have 70 per cent of a successful business than 100 per cent of a failure.

There are several hundred venture capital firms in the UK, many of them specialising in different sectors of industry or different phases of development. When planning an approach to a venture capitalist, those with matching interests should be carefully selected and contact made with no more than two or three in the first instance – this gives you the possibility of altering your approach if necessary. In all cases, the venture capital manager will want to see a detailed business plan containing cash flow forecasts along with details of the management's experience, the product/process and its competitors, the size of the market and other information of this kind. There are many descriptions of how to write a business plan and some good computer programmes. By all means use consultants if it helps, but it is of great importance that the plan should be written by you. You are the one who is going to have to implement the plan and you are going to have to sell it to potential funders. Help with writing a business plan is given in Chapter 2 and in Section 3.5.1.

The usual exit route, or divestment, is the trade sale, but flotations on secondary stock markets are also an option. In some circumstances, sale back to the founders is possible. The route chosen will depend on several factors, such as the track record of the company, the sums involved and the nature of the products. You should take care to discuss with the investor at an early stage in your relationship the likely timing of the divestment and what route the investor might want to take. You should make sure you are comfortable with the proposals.

Venture funds are unfortunately somewhat wary of start-up companies and new technology, but there are some who specialise in these sectors. The British Venture Capital Association (BVCA) in London publishes a directory of its members and their interests, and the entrepreneur will need this to help in selecting funds to approach. Contact information and details of free booklets are given in Appendices 1 and 8. Venture capital associations exist in most countries and the European Venture Capital Association, located in Brussels, represents around 300 venture capital providers throughout Europe, specifically identifying those specialising in early-stage investment.

An alternative to conventional venture capital is the informal sector, the so-called "business angels". Defined as "individuals of high net worth", business angels, who mostly prefer not to be called that, are frequently willing to invest in the seed and start-up stages of a company and they are often prepared to take an active interest in the business. In most cases they invest only in their own localities and in projects they understand, but many operate outside these restrictions.

By their nature, business angels tend to be hard to find, but increasingly networks of such investors are being created. One of the oldest of these, Local Investment Networking Company (LINC) has headquarters in London but operates through regional agents throughout the country. The National Westminster Bank has launched a nationwide Business Angels Service, and has a national network of Technology Business Managers who offer specially designed financial services and practical advice for innovators. The other clearing banks and accountancy practices should also be contacted to check the help they are currently offering. The BVCA's booklet *Sources of business angel capital* lists about 30 such networks.

As a general rule, business angels are willing to provide small sums from a few thousand pounds up to a maximum of about £100,000 per investment and they usually want to take a hands-on position in the company. This may be extremely beneficial if the angel has complementary skills but can lead to difficulties in some cases. As always, the golden rule is, don't go ahead with the deal if you are unhappy with any aspect of it.

Informal investors vary in their approach to a deal. Some are just as rigorous in their investigations as a venture fund; others work on hunches.

3.4.2   The public sector

In the public sector, a considerable range of *grants, loans and other subsidies* is available from local or national government sources and from the European Union.

National and regional sources

The inventor will find that the majority of schemes offer *grants* that contribute a portion of the proposed project costs, normally in the range 40 per cent to 60 per cent. Loans are provided under the government Loan Guarantee Scheme. This is administered by the DTI to provide loans by approved lenders (i.e. banks) to small firms which would be unable to obtain conventional finance because of a lack of security or proven track record. The guarantee covers up to 70 per cent of the value of the loan and is available for amounts between £5K and £100K for periods from two to seven years. Once the company has been trading for more than two years conditions are eased. However the scope is restricted to companies in the manufacturing, construction or service industries. Unfortunately this excludes a large number of inventors. At the national government level grants for innovation are provided by the DTI.

The most appropriate scheme for inventors at the outset of the inventive stage is the DTI's SMART award, especially if they fall into what has been defined earlier as the "category one" lone inventor. SMART awards are made each year to individuals or companies with up to 50 employees for projects to demonstrate the technical and commercial feasibility of a project. SMART offers up to 75 per cent of project costs to a current maximum grant of £45,000. The proposals must be submitted between January and April and no more than 180 awards are made across the country, so competition is fierce.

Other grants for supporting innovation are directed either to taking the invention from the prototype stage to production (e.g. SPUR) or for collaborative R&D involving a university or research institute (LINK Collaborative Research). Newer schemes are developing within the Technology Foresight programme.

Another category of grants are Regional Innovation Grants (RIGs), but they are selective in another sense. These are for small companies with under 50 employees located in "Assisted Areas". A contact with the nearest Local Government Office (LGO) will quickly inform you if you are lucky enough to have the right location! These grants cover 50 per cent of project costs up to a maximum of £25,000.

European Union grants
This is a complex area which is not open to lone inventors. Eligibility qualifications are rigorous, applications are only eligible from legal entities, not individuals, sole applicants are not eligible and partnerships normally must involve at least two entities each from a different member state. In practice most partnerships have a minimum of three partners and four or five is not uncommon. EU CRAFT awards are available across a wide range of technologies and scientific disciplines specifically for the small and medium sized enterprise (SME). However this definition goes up to quite large companies (defined as having under 500 employees and/or a turnover not exceeding ECU 38 million), so the very small company is really a tiddler on the European scale.

CRAFT offers Exploratory Awards of ECU 22,500 to seek partners and to prepare joint R&D projects at a maximum of 75 per cent of costs.

CRAFT Co-operative Awards are for a group of four or more SMEs which wish to nominate a third party in the form of an R&D laboratory to solve a common technical problem. All project costs are supported up to 100 per cent and go to the third party doing the R&D, and grants may be as high as ECU 1 million. It is very unlikely that an inventor will find this type of support applicable. More information can be obtained by applying to the CRAFT national focal point, Beta Technology. Contact details are given in Appendix 1.

Preparing proposals for all EU grants takes considerable time and effort. Timing of submissions is rigid, evaluation can take three to four months after submission, and even if one succeeds in obtaining a grant contract signature takes a further two or three months. Success rates as low as 5 or 10 per cent are not uncommon. In the case of CRAFT exploratory awards which are relatively streamlined no money is handed out until the project has been completed and the report approved.

In conclusion: the inventor at the initial stage of proving an invention is best advised to apply to national schemes such as SMART.

3.5 Raising the funds

We have looked at the various private and public sources of capital; now we suggest how you should go about actually raising the money.

3.5.1 The business plan

Before setting out to raise funds, you should prepare a business plan. In most cases, providers of funds will want to see this, but even where this is not the case, such as when raising money from family or other close sources, it is a useful exercise to clarify in your own mind the details of what you propose to do.

The business plan should contain some or all of the following:

- executive summary
- description of the project
- management CVs
- market information
- the competition
- 3−5 year financial and cash flow forecast.

Depending upon the nature of the project and the exploitation strategy chosen, not all of these sections may be necessary, but the cash-flow forecast and the executive summary should always be included. The latter should summarise, in not more than two pages if possible, the proposal, the money required and its purpose, and the cash flow.

The business plan is also discussed in Section 2.5.1, and a sample plan is sometimes included in the booklets produced by the high street banks.

3.5.2   Making contact

First read Section 3.5.4 on confidentiality! A sample confidentiality agreement is given in Appendix 5.

Whether or not you decide to use an intermediary to assist you in making contact and in the negotiations (see next section), it cannot be emphasised too strongly that you must prepare your story very thoroughly before approaching potential funders. First impressions are very important and it may be the only chance you get!

Choose carefully the people you want to approach and try to ascertain if yours is the kind of project in which they are interested. Don't contact everybody at once; selecting two or three in the first instance allows the possibility of modifying elements of your plan or the project in the light of the reception it receives.

A "cold" telephone call rarely works; try to get an introduction if you can. If this is not possible, write a letter of no more than one page and send it with a copy of the executive summary to the Chief Executive or Manager of the chosen organisation. Use their names; if you don't know them, ring the switchboard and ask. Follow up your letter with a telephone call if you have not received a response within a week or so.

The objective at this stage is to get an interview at which you will have the opportunity to present your project to an executive of the organisation and you must be ready to give a full and detailed discussion of all aspects of your plan, including problem areas. Don't hide the downside; sooner or later they will find it out.

Listen carefully to their criticisms, even if you think they are misplaced; they may be telling you something about your presentation's style or content which you can profitably use next time.

Even when you find an enthusiastic investor, the negotiations can take several months, although business angels are likely to act more quickly than venture capital managers. The latter have to go through all the details with you, assess what you tell them about the technology and the markets, and persuade their investment committees that you are a good bet. For this reason you must start your fundraising efforts in good time.

3.5.3   Using intermediaries

What is an intermediary?

By intermediary we mean any *professional* adviser or consultant. They may be licensing consultants, accountancy practices or business consultants and they may be individuals or firms. To use or not to use such assistance is a question which must at some time be faced by the inventor who has neither the experience nor the time to devote to the time-consuming exercise of raising funds.

A simple rule of thumb on when to bring in an intermediary depends on the amount of finance required. Effective intermediaries can certainly be worth their fee if the sum being sought goes well beyond £100,000. For more modest amounts of capital it does become very questionable whether an intermediary is cost-effective. On the other hand it has been claimed that the time and effort involved in raising £20,000 is no different from raising £2 million. Before making a final decision on this matter, you will need to consider carefully:

*   how to choose an intermediary: where to find one and how to select the right one for the job
*   how the intermediary is to be paid: by up-front fees, payment in kind with equity from the venture, or a contingency arrangement where a percentage of the capital that is raised is paid in fees
*   the need to establish clearly defined milestones that the intermediary must meet on the route to obtaining a financial package
*   the need to ensure that the intermediary understands how to handle confidential matters (see Section 3.5.4)

• the need to ensure that the intermediary obtains all fees from you and is not also obtaining a finder's fee from the source of the capital, which not only causes serious conflicts of interest but automatically makes the financing more expensive for you.

Choosing the intermediary

In the past your accountant or solicitor would be ready to act in this role on your behalf. Today, unless they are registered as financial advisers, they will not be able to act in this capacity. This section assumes that the typical inventor will not have the financial muscle to go to the main source of financial intermediaries such as merchant banks and the appropriate departments of the major accountancy firms. Certainly your accountant will be able to put you in touch with one or more, locally based, financial advisers with a good track record and with experience in arranging finance for company start-ups in manufacturing or some other technology-based venture. Another source of advice will be the nearest office of Business Links, Business Shops, Business Connect, or LEDU, where you should request a meeting with the investment and business counsellor, if this is available.

Now armed with several names your next step is to arrange to meet at least three potential intermediaries. Good preparation on your side is essential for this meeting, including a short description (one page only if possible) of your invention, a copy (if it is available) of the business plan and details of any existing approaches that you may have already made to raise finance. Indicate whether or not these approaches are still active and whether you require the intermediary to take over these contacts on your behalf.

The interview should allow you to establish if the personal chemistry is working well and what the intermediary's experience is in the type of venture you are embarking on. Request an outline of the steps the intermediary will follow in seeking finance and the likely timescale. Finally, if you are satisfied with what you are hearing, ask the intermediary to supply you with a proposal plus a draft letter of engagement that would establish the basis of the services and remuneration. Raising finance is a painstaking process and can easily take three to six months or even longer. It is important that the intermediary you select recognises this and can maintain the services over this period. For this reason, even if you are paying the adviser on a contingency basis (for example, fee paid on successful conclusion of the business, or dependent upon the value of a licence), you will be asked for an initial commitment as an advance on the final fees.

Paying the intermediary
The contingency fee formula that is typically used is based on the Lehman Scale. The figures given here are merely used as a demonstration of how the scale operates:

Amount of capital	% fee	Actual fee
< £250,000	5	£12,500
£250,001 – £500,000	4	£20,000
£500,001 – £750,000	3	£22,500
£750,001 – £1,000,000	2.5	£25,000
> £1,000,000	2	Max. £30,000

Note at the lower end of this scale, the intermediary will be receiving a basic fee irrespective of the capital raised.

Do not accept any proposal where the intermediary offers to work on a daily rate basis, as this is an open-ended commitment you cannot afford to take. On the other hand, with contingency fees it is unwise to squeeze the percentage rate too low, as you risk the intermediary losing interest in your case if there is too little money to be made out of the deal.

Whilst you may be loathe to commit part of your hard won capital for this task, bear in mind that if the service is done effectively you will know that you have done your best for this critical job of providing capital for your finance.

However, there is a final word of warning. The appointment of the intermediary does not absolve you from all the effort needed to raise the capital. To get the best out of the intermediary you must co-operate fully, be ready to attend any meetings with investors, and provide any additional information required promptly. Make the intermediary a member of your team, and if the relationship goes well you will gain a valuable ally for your venture over many more years.

3.5.4   Confidentiality

In the course of describing your project to a potential investor or your financial adviser (assuming one is appointed) details will need to be supplied of the invention, its manufacturing process and the market it serves. The business plan will describe in greater or lesser details all these matters and in fact it may be designated a confidential document.

What is involved is the disclosure to third parties of enough information on your invention to create confidence and enthusiasm without disclosing any commercially or technically sensitive information.

If you have had a patent application published its contents are already public knowledge, but it will include only a part of the know-how being developed to promote the invention. This know-how also needs to be protected from unnecessary disclosure. This process needs careful thought and preparation. A common mistake often made by inventors when meeting those who may help is to refuse to disclose anything. This makes it difficult to convince enquirers of the potential of the invention and they lose interest in the project.

To overcome this problem the inventor can ensure adequate discussion without total and free disclosure in this way:

• list all the confidential information over and above that covered by the patent
• decide which parts of this information may be disclosed, which will be the minimum necessary for understanding and assessment
• draw up a Non-Disclosure Agreement (NDA) which any potential investor who has requested sensitive information will be required to sign.

Some advice on NDAs is appropriate as the authors frequently see agreements so poorly drafted that they provide no protection whatever to the owner of the know-how.

In the NDA define clearly who are the parties to the agreement and what their responsibilities are. As the inventor you must indicate very clearly that the information you are handing over is confidential whether it is a document, drawing, sample or an oral communication, and state the uses that may be made of it. Any confidential document or material remains your property and must be returned to you according to specific conditions.

Define the period of confidentiality, normally three or five years. State who has authorised access to the confidential information and who has not. Ensure that the person who signs the agreement has an authority to sign such agreements. Finally, and most important, keep a record of every meeting where confidential information is passed on to signatories to the NDA and exactly what they received.

These procedures will enable you to challenge any loss of confidentially should it occur. The NDA may not be a total remedy but it should enable you, as the inventor, to recover damages from a guilty party.

A word of comfort: if you have signalled to a third party that you are serious about confidentiality matters most people will respect this and take the necessary precautions. Another comfort is that your invention is in a state of flux and progress. What one day had appeared highly confidential can become, as your know-how progresses, better protected.

A sample confidentiality agreement is given in Appendix 5.

3.6 Managing your investors

Your investors may be family, banks, venture capitalists or business angels and it is their money, corporately or individually, that you are using. It is in their interests to help you to succeed but in turn, you have to help them.

Whether you have one or several investors, it is up to you to keep them happy and this means providing all the information they want, when they want it.

The best way of doing this is to provide management accounts. Investors vary in their approach, but it is usual to expect to have to prepare a complete set of accounts every month, benchmarked against your budget figures. This may sound like a chore but modern software makes it easy, and it is an excellent discipline.

Venture funds often require a seat on the Board for one of their executives. Keep their representative informed on what is going on, the bad things as well as the good. If you have problems, the investors can play a major part in helping you to clear them, especially if they know soon enough.

Remember that in real life your business will start to diverge from the business plan which cost you so much blood, sweat and tears at the beginning. This is normal but having the investors on your side will help you over any bad patches.

From time to time you may need new money in the firm. Whether or not your investors choose to provide this will depend on circumstances but they will certainly help you as much as they can to find it, providing they have confidence in your ability to pull through. So don't hide anything, keep them informed and you will have the best of chances.

3.7   The future

Sometimes your investors, especially if they are family or friends, will stay with you indefinitely, drawing dividends on their shares as your company becomes profitable. More often, professional investors will at some point want to exit from the investment. You should be ready for this and ideally will have discussed it with them even before the investment was made.

Whether the exit is via a trade sale, a flotation, or by buying back their shares, a new ball-game is beginning. You will have to consider your own position in this: do you want to be part of another company? A public company? Or buy back the shares and return to full ownership? It is worth considering these options before you start in business and ensuring that you and the investors see eye to eye.

Be patient. Outside circumstances can affect your progress, sometimes substantially, but if you keep your eye on the ball and do everything correctly, you have a very good chance of succeeding.

3.8 Organisations to approach for advice or services

Your local business information centre should be able to point you to sources of help, and many offer free initial advice sessions with experienced counsellors. Contact numbers for details of your nearest centre are given in Section 8.7.7.

Most of the major high street banks produce free booklets of advice and information for people considering starting up in business. As well as giving details of their particular financial packages they can include very useful checklists of things to think about, and sample business documents. One example of specialised help for innovators is NatWest's Technology Business Support Network.

- The **British Venture Capital Association (BVCA)** publishes free information on sources of UK venture capital. The free guides mentioned in this chapter are listed in the bibliography in Appendix 8.
- The **Institute of Chartered Accountants in England and Wales** has district societies, each with an administrator who can refer enquirers to local firms with experience in the appropriate area.
- **The Institute of Chartered Accountants of Scotland** can provide a list of Chartered Accountants in your local area. The local list is available free from the Institute's Member Services Department.
- **Local Investment Networking Company (LINC)** is a nationwide business introduction service linking private investors (business angels) with businesses seeking equity investment between £10,000 and £250,000. It does this via

the monthly *LINC Bulletin*, Company Presentation Days, and Database Matching Service. LINC has a network of regional offices, and information on the nearest one can be obtained from the London Head Office.

- **The London Enterprise Agency (LEntA)** is Britain's oldest and largest enterprise agency, funded by 19 large companies and the Corporation of London. Since 1979 LEntA has helped over 70,000 businesses set up and expand, and has a specialist consultancy and a range of products to help you get the most from your business.
- **Venture Capital Report (VCR)** matches businesses seeking capital with business angel investors through its monthly *Report*. Innovators pay a fee to appear in the *Report*. VCR also publishes *The Guide to Venture Capital in the UK & Europe*.
- **The European Venture Capital Association (EVCA)** represents over 200 leading venture capital companies in Europe. They produce a yearbook and directory which contains a list of members and information on funds and investment criteria. Contact details are given in the organisations appendix.
- **Innovation Relay Centres**, listed under that heading in Appendix 1, can be contacted for details of European Union support.

Further information on business information and services can be found in Chapter 8, which includes a discussion of the Internet as a source of business information in Section 8.7.8.

Contact details of all the organisations mentioned in this chapter are given in Appendix 1.

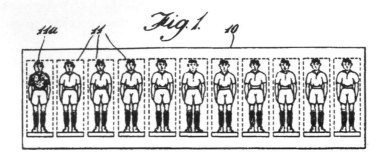

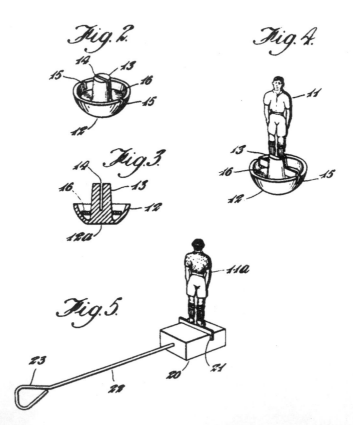

Figure 4. British patent number 616715. Peter Adolph's Subbuteo
table game patent, applied for in 1946.

4
Patents

Jeremy Phillips

4.1 On patents and patent law

This chapter offers a brief introduction to United Kingdom patent law for the non-lawyer, together with some guidance as to the interface of UK and foreign patent protection. While every effort has been made to ensure its accuracy, the chapter encapsulates in generalised form a vast and complex subject. Before taking any major decision with regard to any invention, the reader is urged to seek advice from a qualified professional practitioner in the field of patent law.

Above all, the reader should remember that the accuracy or reliability of any written statement of the law may be influenced by legal developments which transpired after the date of publication. Because the patenting of inventions and their subsequent commercial exploitation is a fast-moving field, the reader should assume that anything which was published more than two years previously is bound to have been superseded by a more recent work on the subject.

This chapter answers the following questions:

- what is a patent?
- what sort of inventions can be patented?
- how long does patent protection last?
- how far does patent protection extend?
- what protection does a patent give?
- who can apply for a patent?
- who is entitled to receive a patent?
- how is a patent obtained?

(12) **UK Patent Application** (19) **GB** (11) **2 303 585** (13) **A**

(43) Date of A Publication 26.02.1997

(21) Application No 9615335.8

(22) Date of Filing 22.07.1996

(30) Priority Data
(31) 9515116 (32) 24.07.1995 (33) GB

(71) Applicant(s)
David Edward Cross
Hedge End, Hurst Road, E.Preston, LITTLEHAMPTON,
West Sussex, BN16 3AP, United Kingdom

(72) Inventor(s)
David Edward Cross

(74) Agent and/or Address for Service
David Edward Cross
Hedge End, Hurst Road, E.Preston, LITTLEHAMPTON,
West Sussex, BN16 3AP, United Kingdom

(51) INT CL⁶
B63C 5/02 15/00

(52) UK CL (Edition O)
B7A AAMM

(56) Documents Cited
GB 2092529 A   US 4759660 A

(58) Field of Search
UK CL (Edition O) B7A AAMM
INT CL⁶ B63C 3/12 5/00 5/02 5/04 15/00
Online:WPI

(54) Boat Support System

(57) A support system for boats 10 has two lateral support frames 1 placed either side of the hull 20 and connected together by a flexible web 2 extending beneath the hull. A tensioning ratchet 3 tightens the web 2 and urges the support frames 1 inwardly against the sides of the hull.

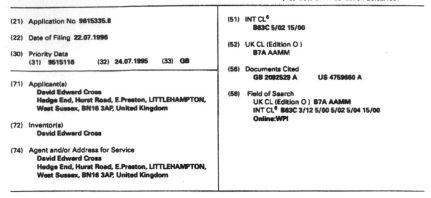

Fig. 1

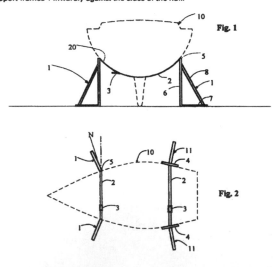

Fig. 2

GB 2 303 585 A

Figure 5. The front page of a published British patent application.

- how is a patent infringed?
- how are patents transferred and licensed?
- how are licensees protected?
- how is protection obtained outside the UK?

4.2 What is a patent?

A patent is a legal grant of an exclusive right in an invention which enables its owner to prevent any other person doing any act in respect of the patented invention which constitutes an infringement of it. This exclusive right is granted for a limited time during which its proprietor may seek to profit from it. Once this exclusive right expires, anyone may make or use the patented invention.

There are two main parts to a granted patent: the *specification* and the *claims*. The specification is a description of the invention which describes it with sufficient detail and clarity to enable a hypothetical "person skilled in the art" to put it into effect. The *claims* are statements as to which parts of the invention are new and inventive and will therefore be infringed by anyone who utilises them. Since most patented inventions relate to very small advances on what was already known, it is vital to disentangle those parts of the invention which are truly inventive from those parts which are already known or used, so that everyone knows what can be done without infringing the patent.

4.2.1 Sources of patent law

In the United Kingdom most of the law which covers the grant of patents is contained in the Patents Act 1977. This Act, which came into force on 1 June 1978, contains the main laws governing how patents are applied for, granted, revoked and infringed. The Patents Act 1977 has been amended from time to time by other Acts of Parliament, of which the most important so far has been the Copyright, Designs and Patents Act 1988.

The detailed operation of the law, dealing with mundane matters such as the forms which are used for applying for a patent or the minimum size of print which may be used in describing one's invention, is contained in subordinate legislation which has the force of law but can be changed much more easily than an Act of Parliament. This is just as well, because routine amendments such as changes in Patent Office fees occur far more frequently than changes in the patent laws themselves. The rules in force at the time of writing are the Patent Rules 1995 (Statutory Instrument 1995 no.2093) and the Patent (Fees) Rules 1996 (Statutory Instrument 1996 no.2972).

Copies of the Patents Acts and the Rules are published by The Stationery Office (formerly HMSO) and may be purchased at The Stationery Office shops. They are also found in many libraries with patent holdings. An inventor who is not also a lawyer is, however, warned not to rely upon his or her own interpretation of statutory provisions – the interpretation of Acts of Parliament is an art in itself, since words often carry definitions which do not accord with their colloquial meaning and since there are also rules (not contained in the Patents Act) which govern the way in which Acts of Parliament are interpreted. Where necessary, professional advice should be sought from a solicitor or a patent agent.

4.3 What sort of inventions can be patented?

Although the Patents Act 1977 lays down what can be patented, it does not define the word "invention". This is because it is assumed that anything which fulfils the criteria of patentability must be an invention.

A useful rule of thumb is that, before an invention can even be considered as an invention under the Patents Act 1977, it must be either a product or a process. In other words it must either be a thing or a way of doing a thing. While being a product or a process is a necessary condition of a patentable invention, it is not however a sufficient condition, as will now be shown.

The Act specifies that certain things are not, for the purposes of the Act, to be regarded as inventions and cannot therefore be the subject of a patent application even if they fulfil the criteria of patentability. In short they are:

(i) discoveries of things which already exist in nature but which were not already known (with the possible exception of human genetic sequences, in respect of which the European Patent Office, which must apply the same law as that of the UK, has indicated that protection may in some circumstances be provided by a patent)

(ii) aesthetic works, such as novels, poems, songs, photographs and sculptures, as well as non-functional designs (these are already protected by copyright or design laws and do not require protection twice over)

(iii) computer programs, which are regarded by the law as if they were literary works and thus protected by copyright on the basis that they are no more than a series of instructions and that a series of instructions is no more an invention when given to a machine than when given to a human. A computer which has been programmed in an inventive way is not, however, excluded from being an invention and can thus be patented. In the United States the patentability of some industrially useful algorithms was established some years ago. While the current state of United States law is somewhat uncertain, it would still appear to allow the patenting of inventions which would not obtain patent protection in Europe.

(iv) the presentation of information, for example a new arrangement of tables of high and low tides which enables users to use them with greater facility

(v) scientific theories, which are not products or processes but merely explain how things are or work

(vi) mental calculations, ways of conducting business and ways of playing games.

4.3.1   The criteria of patentability

Which inventions are patentable? The Patents Act 1977 lays down three criteria which must be fulfilled before a patent may be granted. These criteria, laid down below, are common to most European countries and will, within the next ten years, be adopted almost universally, so an invention patentable in the UK will eventually be equally patentable in most other countries of the world.

4.3.2   Novelty

Every patentable invention must be *novel*, that is to say it must not have been known or available to the public at the time of its "priority date". For this purpose, novelty is a worldwide concept. If an applicant seeks patent protection in the UK on 2 January in respect of an invention which is not known in the UK but has already been filed or disclosed to the public in Murmansk or Montevideo on 1 January, the application must fail.

For this purpose any public disclosure of an invention will defeat a subsequent patent application for it. The disclosure may be contained in a book, an article, a learned paper delivered at a conference – whatever the language – or may be effected simply through the use of the invention in public tests or trials. A thesis deposited in a library to which the public have access will be said to have anticipated the novelty of a subsequent patent application, even though no-one ever took the trouble to read it. The demonstration of an invention at certain very limited types of designated exhibition may not result in a loss of novelty, but expert legal advice should always be sought before demonstrating or exhibiting an invention in any public arena before a patent application has been filed.

The body of available information against which each patent application is measured for novelty is called the "prior art". Prior art also includes a very important body of data which patent applicants will almost certainly know nothing about when they file their patent applications – the contents of the patent applications made by their competitors before the date they filed their own application and which have not themselves yet been published.

Apart from other applicants' unpublished applications, virtually all other types of prior art can be searched before an application is made, to see whether an apparently novel concept has been anticipated by others. Commonly searched areas of prior art include granted patents, published patent applications, papers published in journals or delivered at symposiums, articles in trade journals and exhibition catalogues. Properly conducted prior art searches can be cheap or may be expensive but will almost always yield a good return: if they reveal that an applicant's invention has been anticipated by work done by someone else, they will save that applicant the cost of a patent application. However, a search of the prior art which reveals nothing should be carefully interpreted because not all prior art may have been identified by it. Nonetheless, a properly conducted search which does not reveal the existence of any prior art does at least indicate that, if the applicant files an application and is granted a patent, an attack on that patent which is based on its anticipation by prior art is less likely to occur.

Inventors will frequently find that their inventions are not patentable through lack of novelty on account of their own disclosure of it to others before they thought of filing the patent application. For this reason it is essential to keep an invention secret, or to disclose it only under terms of the strictest confidence, before a patent application is filed. Professional communications with a patent agent, solicitor, bank manager or accountant for the purposes of seeking advice, raising capital or writing a business plan are assumed to be confidential, but those with prospective customers or retailers are obviously not. The way in which the secrecy of an invention can be preserved when disclosing an invention to someone else is discussed in Chapters 3 and 5.

If a patent agent or other professional adviser, or any other person to whom details of an invention is entrusted, discloses information about it by accident or in breach of confidence, a patent application made within six months of the unauthorised disclosure may still be made without loss of novelty. Some countries have more generous provisions which allow a domestic patent application (explained in Section 4.5) to be made without loss of novelty during a "grace period" of six months (e.g. China) or one year (e.g. the United States) following any disclosure by the inventor, whether deliberate or accidental, but it is strongly advised that no disclosure is made which relies on the "grace period" provisions of any country unless prior professional advice is obtained as to the consequences of such reliance.

Once an applicant for a patent has secured a "priority date" for the invention – which is usually the date of the first application for that patent – he or she can disclose it publicly before the patent is published or granted without risking loss of novelty (see Section 4.10.2 on how a priority date may be secured). If an invention is the subject of a patent application which is withdrawn before the Patent Office has published details of the invention, it is not regarded as having been made publicly available and can be the subject of a later application.

4.3.3 Inventive step

Novelty is not the only criterion for the patentability of an invention. It must also possess an "inventive step", which means simply that it must not be obvious to a hypothetical person skilled in the art, a reasonably intelligent but uninventive person who is highly familiar with the prior art and who would be expected to find non-inventive ways of achieving the invention if such were possible.

Inventive step is a difficult concept for laymen to grasp and for the courts to determine. This guide is not the place to criticise it. Readers should however understand that most patent applications and granted patents are vulnerable to an attack on the basis that, despite their novelty, they are not the product of an inventive step. There is no secure means by which this vulnerability may be cured.

For the purpose of determining whether an application possesses an inventive step, recourse is taken to the same prior art as is considered for novelty purposes, subject to one exception: no account is taken of the contents of earlier patent applications which have not yet been published, since there is no means by which the hypothetical person skilled in the art would be able to utilise their technological content.

4.3.4   Industrial applicability

Finally an invention must be "susceptible of industrial application" in any field of use, including agriculture. Certain types of invention are deemed by the law to be incapable of industrial application. They are inventions covering new plants and animals (other than microbiological products and processes for their creation), as well as methods of treatment or diagnosis of the human or animal body. New plant varieties are however the subject of a separate system of legal protection. New animals are not the subject of legal protection.

The patent systems of the United States and some other countries contain provisions which are less restrictive than those found in the UK and other European Patent Convention countries. This means that patent protection may be obtained in those other jurisdictions even if no UK protection is available.

4.4   How long does patent protection last in the United Kingdom?

Patent protection commences from the date of publication of the application for a granted patent, though that protection cannot be enforced by legal means until the date on which notice of the patent grant is given in the *Patents and Designs Journal* of the Patent Office. The publication of the application usually takes place some 18 months after the application is first filed, while the official notice of grant will usually occur between 30 and 42 months after the first filing. The Patent Office now offers an alternative "fast track" route to the grant of a patent, which can lead to grant within a year of filing, but the pros and cons of using this need to be discussed with a patent agent or the Patent Office staff.

A patent which is regularly renewed in accordance with the provisions of the law (see Section 4.4.1) will expire at the end of the twentieth year following its filing date.

In respect of certain categories of patents, notably pharmaceutical products, commercial exploitation is usually delayed for many years while tests are conducted on the safety and efficacy of the patented product. The result of this is that, while a theoretical maximum of 20 years' patent protection is available, the patent owner may have only two or three years at the tail end of the patent's life in which to exploit the commercial monopoly granted by the patent before that monopoly right expires and the invention falls into the public domain, where it may be freely used by anyone. For this reason many jurisdictions, including the member states of the European Patent Convention, the United States and Japan provide for the grant of a supplementary protection certificate which extends the protection of the patent beyond the normal term.

4.4.1   Renewing the patent

The patent, once granted, runs initially for only four years. After this it must be renewed annually. There is a rising scale of fees which must be paid on renewal. Although the Patent Office sends out reminders, it is easy to lose track of renewal dates and deadlines. Many patent owners therefore put the responsibility for the renewal of their patents into the hands of patent agents or commercial renewal services.

Most patents do not live out their maximum term. A small number lapse accidentally through non-renewal. A larger number are deliberately allowed to lapse, usually because the patent relates to an invention which has become technologically obsolete or commercially unprofitable. Sometimes an applicant seeks protection of two or more practical manifestations of an invention, only one of which is worth guarding against competitors. Patents may also be surrendered (Section 4.4.4) or revoked (Section 4.4.5).

4.4.2   Reinstating a lapsed patent

The mere fact that a patent has lapsed does not mean that protection has necessarily ceased, since the lapse of a patent can be cured by payment of a late renewal fee within six months of its expiry. Even after that date, the reinstatement of a patent is possible if the lapse was an excusable one. The position of users of a lapsed patent is difficult: they are safe from infringement actions in respect of the use of a lapsed patent until the date of its reinstatement, but they will be prevented from continuing their use once the patent is back in force. This can cause considerable inconvenience where money and effort have been expended in tooling up, training staff and pursuing sales or distribution agreements.

4.4.4 Surrendering a patent

A patent owner who does not wish to wait until a patent lapses can surrender it. This may be a strategically desirable path to follow where, for example, its proprietor wishes to avoid the expense and inconvenience of revocation proceedings.

4.4.5 Revocation of a patent

Once a patent has been granted, it is presumed to be valid until it expires. This is advantageous to the patent owner, as is the fact that it is not possible to institute legal proceedings in order to halt a patent application in the course of its passage through the application procedure. However, once a patent is granted it is possible for anyone to commence revocation proceedings. If a patent is revoked it is as though it never existed, at least so far as infringement proceedings are concerned. It does not follow in most cases that a patent licensee can reclaim royalties paid for permission to use a patent which has been subsequently revoked: legal advice should always be sought in construing the terms of patent licences if any question of refund of royalties is raised.

4.5 How far does patent protection extend?

In general, each country has its own patent system and grants patents only in respect of its own territory. An application filed in any country's national patent office is called a national, or domestic, application in order to distinguish it from an application filed in a regional or international patent office.

A UK patent, once obtained, may be individually registered with the patent offices of a number of other jurisdictions which do not have their own patent examination systems, for example Hong Kong, Cyprus and some of the island states in the West Indies. On registration the patent will be protected in those countries as though it had been granted there.

There is no such thing as an international or world patent, although the term is often used to denote a patent application filed under the Patent Cooperation Treaty at the International Bureau administered by the World Intellectual Property Organization in Geneva (see Section 4.14.3). There are however four regional patent offices:

(i)   the European Patent Office, which currently processes a single filing which seeks national protection in one or more of the 18 member states of the European Patent Organisation (see Section 4.14.2)

(ii) that of the Organisation Africaine pour la Propriété Industrielle (OAPI) in Yaoundé, which grants a single patent covering all 15 French-speaking African states which belong to that organisation

(iii) that of the African Regional Industrial Property Organisation (ARIPO) in Harare, which processes by means of a single filing an application for national patent protection in one or more of its 14 member states

(iv) that of the Eurasian Patent Office in Moscow, which enables national patent protection to be secured in a number of territories which were formerly members of the USSR.

A current list of countries in which protection may be obtained through the PCT and regional patent arrangements is given in Appendix 7.

4.6 How much does patent protection cost?

There is no set cost for pursuing and attaining a patent – each case depends upon factors which are specific to it, such as the complexity of the technology to be protected, the amount of work involved in drafting the specification and claims, as well as the degree of interaction which is needed with the Patent Office in order to ward off objections or make amendments. UK Patent Office fees at the time of writing are given in Appendix 4.

Patent Office filing fees are relatively low. The main cost of patenting is incurred in retaining the services of a patent agent. It is not necessary for an applicant to use a patent agent and the Patent Office policy is to be particularly helpful to unassisted applicants. However, unless a patent applicant is highly literate, has a good general scientific education, is not in a hurry to obtain a patent and has nearly unlimited time to deal with the application, he or she will be strongly recommended to use a patent agent's services. Advice on obtaining a patent agent is available from their professional body, the Chartered Institute of Patent Agents (CIPA), described at the end of this chapter.

4.7 What protection does a patent give?

Any unauthorised action done in respect of a valid patent and which falls within the scope of that patent's claims will infringe that patent. Where every element of the patent claimed is copied or used, there is said to be a *literal infringement* of the granted patent.

Frequently an alleged infringing act will not be a literal infringement, since it will involve the use of some or most of the elements defined by the patent owner's claims, but with some changes incorporated. Case law on this problem is complex and confusing, but the following crude conclusions can be drawn:

(i) where the alleged infringement involves the substitution of one feature of the patented invention with a mechanical equivalent, infringement will normally be found

(ii)   where the words of the patent claim are taken as indicating a general or functional requirement rather than a literal one, infringement will normally be found (for example if a feature of a patented invention is required to be "vertical", it should not be possible for another person to get round the patent by employing the same feature at a couple of degrees off vertical, unless "vertical" is intended to be a precise and literal term)

(iii) where the change results in a substantially different mode of operation or performance of the invention, infringement will be unlikely to be proven.

4.8   Who can apply for a patent?

Any person may apply for a patent, whether or not the inventor of the invention in question. "Person", for these purposes, means a legal person. This is a broad category which includes humans, incorporated companies and government bodies.

4.9   Who is entitled to receive a patent?

The patent will be granted to whoever applies for it, with just two provisos:

(i)   where another person can show that he or she, and not the applicant, is entitled to be granted the patent by virtue of the fact that he or she is the inventor, the inventor will obtain the patent

(ii)   where another person can show that he or she is entitled to the invention because he or she is the "successor in title" (the subsequent owner) of the invention, he or she will have a better entitlement to the patent than either the patent applicant or the inventor.

4.9.1   Naming the inventor

The inventor is legally entitled to be named as such in the ultimate patent grant. The absence of the inventor's name from the patent application does not however affect the validity of the resulting patent.

Where an invention is made by a team rather than by an individual, it is not uncommon for a single member of the team, or a selection of its members, to be named as inventor. The naming of inventors is frequently a matter of great importance to those concerned, particularly where a promotion or academic credentials may depend upon the choice of named inventor, so sensitivity should be shown in dealing with this matter as early as possible in the application procedure.

4.9.2   Joint inventors

Most major inventions today are the result of the collaboration of two or more individuals, often working as a team. Where the core of an invention − the part of it which is reflected in the patent claims − is the work of more than one person, then the result is probably a joint invention and each of the people so involved should be named as an inventor in the patent application and subsequent grant. Where, however, the inventive core of the claimed invention is the work of just one person, the collaboration or assistance of other people in an essentially uninventive capacity (for example, in conducting tests to verify an experiment or in constructing a prototype) does not turn the invention into a joint invention.

The collaboration of inventors and non-inventive assistants is a matter of great delicacy. Employers often prefer to spread the credit for making an invention as widely as possible as a sign of appreciation of the contributions made by team members, while inventors are frequently resentful of what they see as a dilution or diminution of their inventive effort, particularly if a consequence of it is that they may be deprived of part or all of their entitlement to extra statutory remuneration when a patent has been profitably exploited by the employer (see Section 4.9.4).

4.9.3   Employee inventors

Where an invention is made by an employee in the course of regular or other specified employment duties, and an invention could reasonably be expected to result from the performance of those duties, the invention – and therefore any resulting patent – will belong initially to the employer. The same applies even where the invention is made outside the terms of the employment duties, where the circumstances of the employment would render it inappropriate for anyone other than the employer to own the invention. In all other cases the invention will belong to the inventor. These rules apply in respect of all inventions and not just patentable ones.

Clauses are frequently found in employment contracts which stipulate that all rights in inventions or other intellectual property created by an employee will belong to the employer. The Patents Act 1977, however, renders such clauses unenforceable to the extent that they deprive an employee inventor of any right to which he or she would be entitled under that Act.

Only employees under a contract of employment are governed by the provisions described above. Freelance and consultant contractors who are not employed under an employment contract are not affected by them. This means that, subject to any contract terms to which they give their agreement, they are initially entitled to the ownership of the invention.

4.9.4   Compensation for patents used by employers

There are two circumstances in which an employee inventor may claim further payment from an employer in addition to the pay received for working:

(i)   where the invention belongs initially to the employer because it was made by the employee in the course of employment duties and the patent resulting from the invention confers an outstanding benefit upon the employer and, taking all the circumstances of the invention into consideration, it would be fair for the employee to receive a share of that benefit

(ii) where the invention belongs initially to the employee who sells or licenses it to the employer for a substantial undervalue.

Once the various statutory requirements relating to either of these states of affairs have been met (which in practice is very rarely), the Comptroller of Patents or a court may order the employer to pay one or more sums to the employee if the parties cannot voluntarily agree how much the latter is to receive.

These compensation provisions cannot be excluded by the terms of the contract of employment but, in the case of (i) above, they can be supplanted by the terms of a collective agreement made by the employer with a trade union which governs inventions of the type actually made, where they have been made by employees of the category to which the inventor belongs. There are hardly any collective agreements of this nature, which makes this legal provision of little relevance today.

4.10 How is a patent obtained?

An application must be made to the Patent Office. The Patent Office will examine it to make sure that it complies with certain application formalities. It will then search the prior art in order to identify earlier patents which may affect its patentability. At 18 months from the priority filing (explained in Section 4.10.2), the Patent Office will publish the application, following which a substantive examination of novelty and inventive step will take place. If the substantive examination is successfully negotiated, the application will then proceed to grant.

4.10.1   The application

An application will only be accepted by the Patent Office if it is made on the right form. Copies of the form are available from the Patent Office and from patent agents. Contrary to the notion popularly portrayed by cartoonists, it is neither necessary nor permitted to deposit a model of an invention at the Patent Office by way of application.

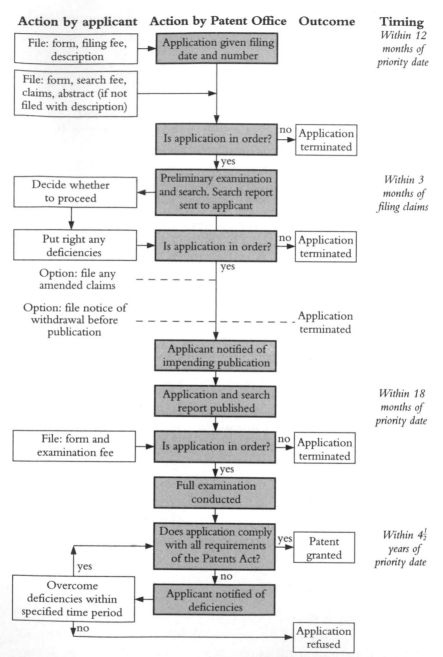

Action by applicant	Action by Patent Office	Outcome	Timing

Action by applicant

File: form, filing fee, description

File: form, search fee, claims, abstract (if not filed with description)

Decide whether to proceed

Put right any deficiencies

Option: file any amended claims

Option: file notice of withdrawal before publication

File: form and examination fee

Overcome deficiencies within specified time period

Action by Patent Office

Application given filing date and number

Is application in order?

Preliminary examination and search. Search report sent to applicant

Is application in order?

Applicant notified of impending publication

Application and search report published

Is application in order?

Full examination conducted

Does application comply with all requirements of the Patents Act?

Applicant notified of deficiencies

Outcome

Application terminated

Application terminated

Application terminated

Application terminated

Patent granted

Application refused

Timing

Within 12 months of priority date

Within 3 months of filing claims

Within 18 months of priority date

Within 4½ years of priority date

Note: *Optional accelerated procedures including combined search and examination are possible.*

Figure 6. Simplified flowchart of an application for a UK patent.

The application must be accompanied by a brief abstract of the invention, which has no legal effect but acts as a helpful guide to the Patent Office as to the expertise required by the patent examiner who will deal with it. The application must also be accompanied by the filing fee.

Although a UK resident may wish to obtain a foreign patent, the Patents Act 1977 makes it a criminal offence to do so before six weeks after filing a UK application or before being given permission by the Comptroller of the UK Patent Office. This gives the Ministry of Defence a brief window of opportunity in which to identify militarily sensitive or valuable technologies which they may wish to suppress or appropriate. Once the period of six weeks has elapsed, the applicant may, if wished, file a foreign or international patent application. If the applicant does not wish the UK application to proceed, it can be withdrawn or abandoned.

A patent application is as much a piece of intellectual property as a granted patent. It may be bought or sold and the patent which has not yet been granted can be licensed to a third party.

4.10.2   Obtaining a priority date

The priority date is normally the date upon which a patent application is received by the applicant's national patent office. The priority date of an earlier patent application can be claimed if the subsequent patent application can be said to be fairly based upon the earlier one. In the case of foreign applications, the provisions of the Paris Convention (see Section 4.14.1) give applicants from convention countries a period of 12 months from the date of their first filing in which they can file a patent application in any other convention country and still enjoy the priority date of their first application. Where a patent application is filed and subsequently withdrawn before publication, it may be filed again on a later occasion but will not be able to enjoy the priority date of the earlier filing.

It is possible in Britain to obtain a filing date by filing even a provisional statement of an intention to apply for a patent, so long as this filing is completed and turned into a full application within 12 months from the date of the provisional statement. Such a statement will secure an early priority date if it includes:

(i)   a statement that a patent is to be sought

(ii)   a description of the invention upon which the later filing of the specification must be clearly based

(iii)   an indication as to who is seeking the patent

(iv)   the relevant filing fee.

   This provisional filing can be withdrawn at any time before the 12 months' period passes for its conversion into a full application. If it is not withdrawn or completed, it is of no effect and cannot be used as the basis of a priority claim in respect of a subsequent application. If withdrawn it can be resubmitted at a later date.

   The decision when to file – and thus obtain a priority date – is a difficult strategic one which should be based upon firm commercial and industrial advice if it is available. The earlier a priority date is, the less likely the applicant is to find that the application has been anticipated by any competitors. However, the later a priority date is, the longer into the future will potential protection be available. In a crowded, fast-developing field of technology such as consumer electronics, it is usually wise to file as swiftly as possible in order to keep ahead of the market. In a slow-moving field where the market is not led by continued demand for new technology, such as civil engineering of flood barriers, there is less urgency to race to the Patent Office and a later application may prove beneficial in the long run.

4.10.3   The preliminary examination

Once an application has been received by the Patent Office, it is subjected to a brief preliminary examination which enables the Patent Office to ensure that the application accords with legal requirements and that it is accompanied by the relevant abstract and filing fee. Errors and omissions may be remedied, but the priority date will only run from the first date upon which the application is complete, unless an earlier priority date has already been obtained by a previous application or a provisional application.

4.10.4 The search

The Patent Office then conducts a search of relevant prior art which it considers appropriate for the patent applicant to consider when deciding whether to proceed with the application. Although the applicant may well have conducted or commissioned a prior art search before launching the application, the Patent Office may well cite prior art which was not found, or known but not considered relevant.

The prior art unearthed in the Patent Office search should be the subject of careful reflection on the part of the applicant because it gives a good idea of the obstacles which may have to be overcome in pressing on with an application. Citations are graded according to whether they merely provide background information about the prior art or actually impinge upon the inventive core of the invention in question. The longer the list, the greater the likelihood that the applicant will need to amend or redraft the application in order to steer clear of the obstacles, to drop claims which cannot be sustained or to withdraw the application altogether.

4.10.5 The publication

Around 18 months after filing, the application is published so that it can be scrutinised by competitors and potential objectors to the grant. Figure 5 (page 68) shows a sample front page of a published application document.

Copies of published patent applications may be purchased individually from the Patent Office or from the British Library's patent document delivery service, Patent Express.

4.10.6 The substantive examination

Upon receipt of a request from the applicant, the Patent Office will conduct a thorough substantive examination of the novelty and non-obviousness of the invention. If this examination is successfully negotiated, the application will then proceed automatically to grant.

4.10.7   The grant

A notice of the grant is published in the *Patents and Designs Journal*, following which it becomes possible for the patent owner to sue infringers. The grant is also recorded in a register, which gives details of its ownership, inventorship and sundry other matters such as transfers of title and licence details. The register is available for public inspection, upon payment of the relevant fees. The patent application is republished on grant, and Chapter 7 discusses how to access information on patents which have been published.

4.10.8   Post-grant opposition

Prior to the Patents Act 1977 it was possible to lodge an opposition to a patent application, which could be contested through the courts. This tended to delay the patent grant in many cases, also resulting in duplication of effort and expense in that the same grounds upon which a patent application could be opposed were the grounds upon which the granted patent could be revoked. Accordingly the right to institute legal proceedings in order to challenge a pre-grant application was abolished. In the UK, therefore, there are no longer opposition proceedings but only revocation proceedings. With respect to patents granted under the European Patent Convention, however, the post-grant revocation proceedings before the European Patent Office are rather confusingly termed opposition proceedings.

While it is now only possible to oppose a patent after it has been granted, there is a mechanism whereby any party can write to the Patent Office and state the ground(s) upon which it is believed the application should not be granted. The Comptroller is supposed to take these observations into account but cannot be forced to take any further steps. If it is believed that no notice has been taken of information submitted a judicial review may direct notice is taken, but cannot compel a rejection of the application.

The procedure is little used, and, where it is, rarely results in an application being rejected.

4.11 How is a patent infringed?

The whole point of having a patent is that it gives its owner the power to prohibit (or more often to license for money or some other advantage) the doing of certain acts in which the patent gives its owner a monopoly. If a patent cannot stop such acts being done, the patent is of little or no value. It is therefore by reference to the legal definition of infringement that the patent derives its strength. There are two types of infringement under the UK law: they are often called primary infringement and secondary infringement for ease of reference by lawyers and they are described under those headings below.

4.11.1 Primary infringement

The patent monopoly gives its owner the exclusive right to stop anyone else making, selling, offering for sale, possessing and disposing of patented products and products made by patented processes, as well as using patented processes. The importation of goods into the UK can also be prevented where, if those goods were made in the UK, their making would have constituted an infringement there.

This patent monopoly is what is often called an absolute monopoly. This means that a person infringes a patent by doing the statutory infringing acts even if not aware that those acts infringe and even if that person has no idea that the patent exists.

4.11.2 Secondary infringement

Further infringing acts are committed by the manufacture or supply of non-infringing products if they are to be used for the purpose of putting the invention into effect.

4.11.3   Defences to an infringement action

The Patents Act 1977 provides a small list of defences to an action for patent infringement. This list ensures that the contemporaneous manufacture of a patent product in fulfilment of the requirements of a medical prescription will not constitute an infringement, that testing out a patent to see if (or how) it works is permitted and that aeroplanes and ships which are merely in transit over or around the UK and which contain infringing parts are not taken to be infringements of a UK patent.

4.11.4   Suing for infringement

Where a patent owner wishes to sue an infringer, the normal course is to initiate proceedings before a court. There are two courts in the UK which specialise in patent matters: the Patents Court and the Patents County Court. The Patents Court is part of the High Court; its procedures are highly formal and litigation before it is extremely expensive. The Patents County Court aims to provide a cheaper, swifter and less procedurally complex forum for the resolution of patent disputes, relying more on written submissions and less on oral testimony.

Where a patent owner is suing a licensee, the patent owner may be able to do so by means of an arbitration if an arbitration clause has been incorporated into the licence. The arbitration clause may require that disputes be submitted to an independent expert with scientific or technical knowledge, to a lawyer or to a combination of lawyers and experts. Further information on arbitration and alternative dispute resolution is available from the Arbitration Center of the World Intellectual Property Organization in Geneva.

4.11.5   Remedies for infringement

In an appropriate case an award may consist of one or more of the following remedies:

(i)   an injunction. This is an order to the infringer to stop infringing acts. It is a very powerful order, since a party who disobeys it is in contempt of court and may face imprisonment or a heavy fine. Injunctions may be *interlocutory* (to take effect at an early stage in the dispute, so that the patent owner's position is not commercially damaged by continued allegedly infringing acts before the full trial takes place) or *final*. An injunction will not automatically be granted: it is awarded only at the discretion of the court and in the light of the conduct of the parties and other circumstances.

(ii) damages. The courts may award a compensatory sum of money, which need not be sterling, by putting the patent owner into the position he or she would have been in if the patent had not been infringed. Where a patent has been licensed, the courts will usually take note of the licence rate when fixing the amount of damages while, in cases where the patent has not been licensed, the courts will be influenced by considerations such as the actual or presumed loss of sales or contracts resulting from the infringement.

(iii) an account of profits. This is a sum of money which does not represent the damage done to the patent owner (which may be quite small) so much as the profit enjoyed by the infringer (which may be very large). An account may only be awarded instead of damages, not in addition to them.

(iv) delivery up or destruction of infringing products. This is a discretionary remedy, which the courts will only order when they think it fair in all the circumstances.

(v) a declaration that the patent is valid and has been infringed. This is a useful remedy where the patent owner simply wants to establish the principle of the matter, leaving the subsequent handling of the matter to negotiation (for example by turning an infringer into a licensee).

4.12 How are patents transferred and licensed?

Patents may be bought and sold, mortgaged, leased and held by way of security, so long as the relevant provisions of UK law and the Patents Act 1977 are adhered to. Some transactions in patents must be recorded in the Register of Patents, a record which is open to public inspection.

An assignment of a patent must be made in writing and must be signed by both parties. Where an invention is made by employees in the course of their employment and belongs automatically to their employer (see Section 4.9.3), no such assignment is necessary under UK law – but it is often advisable to have clear documentary evidence, such as a signed agreement confirming that the invention belongs to the employer, before embarking upon litigation in any jurisdiction which is unfamiliar with UK employee invention provisions.

4.12.1   Different types of licences

An invention for which a patent has been granted, or indeed sought, may be used by someone other than the invention's proprietor, but only with the proprietor's permission or *licence*. The patent proprietor in this case is known as a *licensor*, while the person who uses the invention is known as the *licensee*. (The commercial decision to license or transfer intellectual property is discussed in Chapters 2 and 3.)

As explained in Chapter 2, licences may be exclusive, sole or non-exclusive. An exclusive licence is one which is granted to just one person, who may use the invention to the exclusion of everyone – even the licensor. A sole licence is one which is granted to just one person, who may use the invention in addition to its owner. A non-exclusive (or simple) licence is one which may be given to any number of licensees, who effectively compete with each other while making or using the invention.

A licence may be granted for the full term of the patent, or for any shorter period. It may not however be granted for a longer period. It may be granted for some of the rights covered by the patent, or for all of them. Most patent licences also permit the licensee to use any know-how, trade secrets or manufacturing techniques which are associated with the invention.

Most licences are voluntary, which means that they are contracts freely negotiated between the owner of the invention and the person who wishes to use it. The law provides however for two other species of licence – the licence of right and the compulsory licence. The various types of licence are outlined below.

4.12.2   Voluntary licences

Because it is freely negotiated between its parties, a voluntary licence is governed by the law of contract, a mix of statute and case law which determines when and whether an agreement is legally binding. Licences involving international trade within the European Union, or which affect the market within a substantial area of the European Union, must also conform to the terms of the Technology Transfer Regulation if they are to avoid any risk of contravening European competition law. Since both contract law and European competition law are subjects of some complexity, it is strongly recommended that the advice of a lawyer is sought before concluding any agreement by which one intends to be bound.

4.12.3   Licences of right

A patent may, at any time during its term of validity, be recorded on the patent register as a "licence of right" patent. This means that any prospective user of the patent may become a licensee on terms which, if not agreed with the licensor, will be imposed by the Comptroller of Patents. A further advantage to the licensor is that, for as long as the patent is so recorded, the renewal fees are only half the normal fee.

If the patent owner wishes to regain the facility of monopolising the use of the patent, the "licence of right" endorsement can be removed, subject to any terms which may be required in order to protect the interests of any licensees who have incurred expense or inconvenience in reliance upon the licence of right.

4.12.4   Compulsory licences

Three years after a patent has been granted but not been used
("worked"), any interested party can apply for permission to use
it, even if its owner is unwilling to grant a licence, if the non-use is
detrimental to one of a number of specified instances of public
interest. This provision is often threatened but rarely invoked,
partly because it is difficult for an applicant for a compulsory
licence to satisfy the requirements for it to be granted, partly
because the threat is often sufficient to persuade the patent
proprietor to grant a voluntary licence.

Special provisions in the Patents Act 1977 enable the
government to make use of patented inventions for health reasons
and at times of national emergency.

In all cases where a compulsory licence is granted or taken, the
patent owner is entitled to receive adequate compensation for the
use of the patent. Where the government uses an invention which
may already be manufactured under a licence, the licensee is also
entitled to seek compensation.

4.13 The abuse of patent rights

The power of the patent right is so great that the law imposes
limits upon its use. First, as mentioned above, the licensee is
protected from having to keep paying royalties even after the
patent has expired. Secondly, the licensee is protected from having
to use certain suppliers of goods or materials unconnected with the
substance of the patent itself (or from being prohibited from using
certain suppliers).

If a patent licence is granted which contains terms forcing the
licensee to use (or not to use) specified suppliers, the result can be
catastrophic for the patent owner – who may be unable
successfully to sue any other person for patent infringement.

4.13.1   Making threats

A patent owner must take great care when threatening others,
particularly retailers of goods, with patent infringement actions

which are unjustified. This is because unwarranted threats are themselves legal wrongs. A party who is the victim of unjustified threats may bring an action for damages in respect of the threats received.

4.14 How is protection obtained outside the UK?

It is possible to apply for patent protection in every country in the world which operates a patent system. This would be tedious and very expensive, as well as fraught with legal and technical difficulties. A number of international arrangements exist which assist the invention's owner in obtaining protection in countries other than his or her own. These are discussed in the following paragraphs.

When most people refer to "international patent protection", they usually mean protection in the 20 to 30 most important countries for them. There is no such thing as an international patent. It is however expected that the trend towards the establishment of an international patent will gather momentum after the turn of the millennium, as more countries adhere to the Patent Cooperation Treaty and the continued effect of the GATT TRIPs Agreement (see page 97) mean that different countries increasingly adopt similar patent systems.

4.14.1   The Paris Convention

Concluded in 1883, the Paris Convention on the Protection of Industrial Property requires its constituent member states, of whom there are now 137, to do two very important things. The first is to agree to treat applicants from other member states as favourably as they treat their own. The second is to require each member state to give a patent applicant 12 months' grace, starting from the date of the first patent application in a Paris Convention country, in which he or she can file a patent application in any other member state while still enjoying the original priority date (see Sections 4.3.2 and 4.10.2). Current signatories to the Paris Convention are listed in Appendix 7.

4.14.2 European patents

Under the European Patent Convention (EPC) 18 member states and 4 associates to date have agreed that a single patent application filed at the European Patent Office (EPO) in Munich will be examined and granted or rejected on behalf of all or any of the EPC countries as an alternative to the filing of separate patent applications in each country. EPO fees are calculated on the basis that it will normally be cheaper to file in Munich if patents are sought in 3 or more EPC countries. Current signatories to the European Patent Convention are listed in Appendix 7.

An EPC application involves putting all one's eggs in one basket: a single decision will involve the grant or rejection of the applications in each country for which protection is sought. Separate national applications may result in a patent being granted in some countries but not others.

An EPC application will generally require the services of a European Patent Attorney. Most UK patent agents are also European Patent Attorneys, who have studied the relevant law and practice and have passed the professional examination.

4.14.3 The Patent Cooperation Treaty

The World Intellectual Property Organization (WIPO) administers the Patent Cooperation Treaty (PCT), under which a single application is received, given a formal examination and then forwarded to as many of the national patent offices as the patent applicant requires. There are, at the time of writing, 83 PCT member states. The PCT results in substantial administrative savings, as well as deferring the costs of obtaining translations of patent applications. It is thus highly popular. Since the European Patent Office has an operative interface with the WIPO's International Patent Bureau, it is possible for PCT applicants to request that their applications be forwarded to the EPO – or indeed to both or either of the African regional patent offices, the OAPI and the ARIPO. Current members of these bodies and signatories to the PCT are listed in Appendix 7.

4.14.4 TRIPs

TRIPs is the Agreement on the Trade Related Aspects of Intellectual Property Rights which came into force on 1 January 1996 under the watchful eye of the World Trade Organization. It is not possible for a patent to be applied for or granted through TRIPs, but this agreement has sought to impose upon all member states an obligation to apply the same criteria in deciding what is patentable and what may be excluded from patentability.

4.15 Professional organisations

- The **Patent Office** is the government agency which operates the patenting system in the UK. Its officials examine and publish applications, grant patents, and provide information on all aspects of the patenting system. The free booklets available should be read by anyone interested in taking out patent protection. The Patent Office has also produced a self-help interactive learning system on laser discs, *The Patent Training Package*, which is freely available to the public in all Patents Information Libraries, and in many university libraries.
- The **Chartered Institute of Patent Agents (CIPA)** is the professional body for patent agents (also known as attorneys) in the UK, and maintains an official Register of Patent Agents. The Institute offers a free information pack on protecting intellectual property, and maintains a regional directory of firms of patent agents.
- The **Institute of International Licensing Practitioners (IILP)** is a body founded to maintain professional standards of consultancy in licensing and related technology transfer activities. For a small fee the Institute will circulate details of products or technology for which licensing is sought to their members, who then contact the enquirer directly.
- The **Licensing Executives Society Britain and Ireland (LES)** is a body which brings together people with diverse interests in the area of transfer technology. Their secretariat

welcomes enquiries about the Society's activities, and has a list of members who can give advice.

*   The **Law Society of Scotland** can provide enquirers with the names of solicitors in their locality who deal with intellectual property (although without recommendation).
*   The **Law Society (England)** and the **Law Society of Northern Ireland** do not provide enquirers with the names of solicitors; for these areas names can be found in local telephone directories, and lists are sometimes available at public libraries and other information points.

Further information on access to patents and information about patents can be found in Chapter 7.

Contact details of all the organisations mentioned in this chapter are given in Appendix 1.

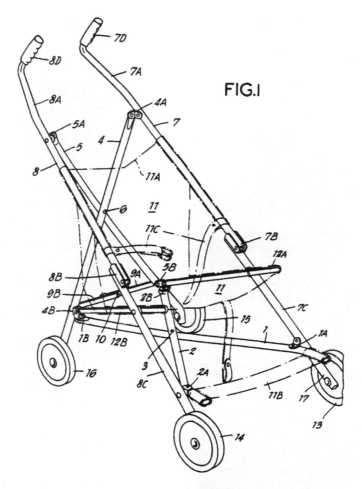

FIG.I

Figure 7. British patent number 1154362. Owen McLaren's folding pushchair patent, applied for in 1965.

5
Protecting Trade Marks, Designs, and other Forms of Intellectual Property

Jeremy Phillips

When someone creates a new design, brand name, invention, literary or artistic work the result is a piece of "intellectual property". The previous chapter discusses the patent system which protects the inventor's rights in an innovative product or process; this chapter deals with the other kinds of intellectual property of importance to inventors.

This chapter covers:

- trade secrets and confidential information as legal subject matter
- utility models and secondary patent protection
- copyright
- registered designs
- unregistered designs
- European design reform proposals
- trade marks.

5.1 Trade secrets and confidential information as legal subject matter

The law of the UK, like that of almost every other industrial country, provides some protection against the unauthorised disclosure or use of confidential information. This is so, regardless of whether the information is the result of inventive activity (such as an invention in respect of which a patent application has not yet been published) or whether it is the result of patient but entirely uninventive labour (such as a company's list of customers or suppliers).

The legal protection of confidential information arises from a series of legal cases which have evolved into a set of precedents that guide our courts in future cases. There is no statute which provides for trade secrets to be protected in the same way that patents, copyrights or trade marks are protected. However, numerous statutes make specific provision for the protection of secret information in certain circumstances. For example the Official Secrets Act 1911 prevents the unauthorised disclosure of any information pertaining to the State, while the Data Protection Act 1984 specifies the circumstances in which personal data held in an electronic data retrieval system should be made available to the people about whom the data is held.

5.1.1   Confidential relationships

Unlike a patent, copyright or other statutory right, a trade secret or item of confidential information is not deemed to be "property" in its own right. It exists only as a force which governs dealings between people. Accordingly, the law will not generally protect a commercially valuable secret unless it can be shown that there is some sort of relationship between the person who wishes to prevent the disclosure or use of the information and the person whom the former wishes to prevent from using or disclosing it.

5.1.2   Relationships with professional advisers

A professional relationship between a patent agent, lawyer, accountant, doctor or bank manager and a client will automatically establish a duty of confidence, by virtue of the trust placed in that person by the client. This is called a fiduciary relationship. Within such a relationship a person can safely disclose information without the need to stress its confidential nature. Indeed, if this were not the case, it would be difficult to imagine how those professional callings could be practised.

5.1.3 Relationships between employers and employees

Employees will generally be under an implicit duty not to divulge or use confidential information to which they gain access by virtue of their employment, even if their contract of employment is silent. In every contract of employment there is an implied "duty of fidelity". This duty is much higher in respect of high-ranking employees than it is for those lower down the company hierarchy, but it can usually be invoked to stop any breach of confidence.

5.1.4 Relationships with independent contractors and consultants

In the absence of any specific provision which insists on an independent contractor respecting the confidentiality of information communicated by a client, the extent (if any) to which a duty of confidence arises is entirely within the realm of the court's discretion in deciding whether an implied duty of secrecy exists. Clearly this duty is more likely to be implied where a software house has been commissioned to write a bespoke credit monitoring program than where a local builder is hired to service the central heating system. To avoid any doubt, an express contract term should always be used when sensitive and valuable information may be exposed to those who are not ultimately entitled to use or benefit from it.

The position of consultants is much the same as that of independent contractors, not least because there is no profession of "consultant" and the fact that such a word is used to describe a contracting party carries no implication as to the legal obligations which arise from a contract made with the consultant.

5.1.5 Protecting confidential information through contract

If any information is valuable enough to need protection, always seek the advice of a professional expert in drafting a set of contract terms adequately to safeguard the information. This advice will generally cost far less than the cost of a patent and can provide both peace of mind and commercial security. Bear in mind that, under English law, contracts are binding once they are made, unless an agreement is expressed to be "subject to contract" (which means "subject to the parties entering a binding contract which will embody the substance of this non-binding agreement"). If you jot down your own home-drafted terms on the back of an envelope and intend to obtain legal advice later, it is quite possible that you will find that you are too late – the contract will already have been made.

5.1.6 Offering secrets to other people

When offering to disclose anything confidential you should try to get a disclosure agreement signed first. A sample disclosure letter might look something like this:

> Dear ...
>
> I write with regard to my disclosure to you of information concerning ....
>
> This information is to be disclosed to you on the understanding that it is confidential and that it must not be disclosed by you to any other person for any purpose except as may be agreed in writing between us. It is also understood that the information may not be used by you for any purpose except as may be agreed in writing between us.
>
> Please find enclosed a copy of this letter, which should be countersigned by you and returned to me as confirmation that you agree to the disclosure taking place on these terms.

A sample of a more formal disclosure agreement for use with a company is given in Appendix 5. Large companies with their own research and development operations are usually reluctant to sign disclosure agreements, however, since they do not want to promise to refrain from using your invention except on commercial terms only to find that one of their subsidiaries had already conceived the same invention. Sometimes it may be possible to persuade the intended recipient of your information that you will disclose it to a neutral third party – perhaps a nominated patent agent – who will then ascertain whether that information already exists within the intended recipient's intellectual property portfolio.

5.2   Utility models and secondary patent protection

In many countries – but not in the UK – an invention which fails to fulfil the criteria of patentability may still enjoy a lower level of legal protection for a shorter term. This protection is sometimes called petty patent protection; in other jurisdictions it may be referred to as utility model protection. The German *Gebrauchsmuster* is the best-known example of such protection in Europe.

Each country's secondary patent protection is unique, since there are no internationally established norms for it. In general, though, such protection is granted on the strength of a deposit of a description or specification which is not examined for novelty or anything else unless an examination is requested prior to an infringement trial. In some countries a patent application may be converted to a utility model application, or it may be possible to seek both simultaneously. Where novelty is required, it may be local rather than global.

At present there are no firm proposals afoot for establishing international norms of secondary patent protection, or even for harmonising the position in Europe where the availability of second tier protection in some countries but not others is certainly capable of distorting the integrity of the single barrier-free internal market.

5.2.1   Which countries offer secondary protection?

Your patent agent or lawyer should be able to tell you if second tier protection exists in a country in which you wish to protect or exploit your invention. Each year the WIPO publication *Industrial property statistics* lists all the countries offering such protection, and gives some idea how many utility model and petty patent applications and grants are made. The statistics reveal that second tier protection is almost always sought and obtained by domestic businesses and applicants, but scarcely ever by foreigners. This itself is an indication that local inventors regard this variety of protection as being of substantial value.

5.3   Copyright

5.3.1   What is copyright?

Copyright is not a single legal right but many quite distinct rights, which may each be held by different people. At the core of each of these rights is however a common thread: the owner of a right which is protected by copyright law has the right to stop anyone else doing certain acts (called "unauthorised acts" by the law) to any work which is copyright-protected.

5.3.2   Where does copyright law come from?

In the United Kingdom copyright is governed by Part I of the Copyright, Designs and Patents Act 1988, as amended from time to time by subsequent statutes. The rights of performers and recording companies to stop the unauthorised use of performances are not strictly speaking a part of copyright, but are protected under Part II of the same Act. The rights of designers in works which form the basis of industrial manufacture but which are not registered under the Registered Designs Act 1949 are protected by unregistered design right under Part III of the same Act.

The 1988 Act came into force for most purposes on 1 August 1989 and governs all works created on or after that date. Works created before that date owe their protection to earlier statutes. In addition to the 1988 Act, copyright is governed by a large body of what lawyers call secondary legislation – the Regulations laid before Parliament and taken to be approved by it if not actually objected to. These Regulations ("statutory instruments") flesh out much of the detail which Parliament does not have the resources to consider in detail, for example the terms under which one library may make a copy of a work which it has in its own stock but which another library wishes to receive for archival purposes.

Guidance as to what may be done without infringing copyright is frequently issued by organisations which represent the interest of rights owners or the information-based professions. This guidance, which may appear in the form of stand-alone guide books or in notices placed near photocopiers, may be an accurate representation of the law but has no legal force and will frequently paraphrase the legal position or seek to apply it in a way which is advantageous to the members of the organisation which has issued it.

5.3.3 What works are protected?

The following types of work are protected by copyright: literary and dramatic works, artistic works, musical works, audio and video recordings, broadcasts, cable transmissions and published editions of works.

The category of literary works is the most important because it is the most pervasive. In addition to any written and printed works, it also includes computer programs (and preparatory materials for them), tables and compilations. Artistic works are also important because that category includes maps, plans, diagrams, production sketches and moulds.

Some things are not protected by copyright, despite the popular misconception that they are. This category includes facts and ideas. While copyright will protect the way in which a particular fact or idea is written or recorded, it will not protect the underlying fact or idea itself.

5.3.4 The copyright monopoly

The monopoly granted by copyright law is a weak monopoly, in the sense that it only protects the copyright owner against the doing of unauthorised acts in relation to the copyright work itself. This means, for example, that copying a computer program infringes copyright in it, but if anyone independently conceives the same program, the person who first created the program cannot complain that copyright has been infringed. Patent rights are much stronger in this regard, since they enable their owner to take action even against a competitor who has no knowledge of the existence of the patentable invention.

The copyright monopoly is of far greater duration than the patent monopoly. Ordinary works of authors and creators (including computer programs, most films, tables, diagrams and compilations) are protected until the end of the seventieth year following the year in which their author died, while sound recordings, films made without direct human authorship, broadcasts and cable transmissions are protected for 50 years and published editions of works are protected for 25 years.

5.3.5 Copyright: general observations

It is frequently assumed that one has to take some positive step in order to obtain copyright protection. This is not so. A work does not have to be "copyrighted" since, under UK law and the law of virtually every country in the world, it is protected under copyright law by virtue of the fact that it has been created and recorded or fixed in some permanent form.

If it should be necessary to prove that one is the author of any copyright work, all that it is necessary to do is to send a copy of the work to one's lawyer, accountant, bank manager or other trusted adviser, taking the precaution of using the recorded delivery service. The copy, once received, should not be opened but should be stored as evidence that, on the date of postage, the author had indeed written the work.

UK copyright law has no requirement that a work be marked with a '©' sign. If, however, any published work bears the '©' sign together with the copyright owner's name and the year of publication, this notice is taken as being presumptive evidence in every country which is signatory to the Universal Copyright Convention that any formalities required for the attraction of copyright protection have indeed been complied with. The "©" sign also warns potential infringers that a copyright has been asserted and it may possess positive advantages if litigation is planned in the United States.

The requirement that copies of published works be deposited with the British Library and, on request, with various other institutions, stands outside copyright law. This means that, even if no copy is deposited, copyright will still be enjoyed by the work's proprietor.

5.3.6   Copyright and international law

The Copyright, Designs and Patents Act 1988 protects copyright works which originate in the UK, whether they are published there or are written by UK authors. The protection of the Act is also extended to various colonial jurisdictions and is applied to a very large number of countries which are signatories of international conventions.

The most important of these conventions is the Berne Convention of 1886, to which the UK and almost every significant industrial country belong. Under the Berne Convention each signatory country has to provide at least a minimum level of protection to its own copyright owners and must accord protection to the copyright owners of other signatory states. Other conventions deal specifically with recorded works, performances, broadcasts and the topography of semiconductor chips.

The World Intellectual Property Organization lists countries which are currently signatory to each convention in the January issue of its journal *Industrial property and copyright*, and a list of signatories to the Berne Convention of 1886 appears in Appendix 7.

5.4 Registered designs

Even where a new product is not patentable, it may possess original features of aesthetic or distinctive design which are themselves capable of attracting legal protection against unauthorised copying or exploitation by others. There are currently two schemes of protection for designs in the United Kingdom: registered design protection in respect of those aspects of a product's design which are judged as being of appeal to the human eye, and unregistered design right for those aspects of a product's design which may lack aesthetic appeal but which are nonetheless of a distinctive character (discussed in Section 5.5). The reader should be warned that design protection is an extraordinarily complex area of law and that expert professional advice should always be taken, either when seeking to protect a design feature against copying or when proposing to incorporate a feature of another's design into one's own work.

In addition to design protection under UK law, European Union proposals for two Community-wide rights are also currently under discussion. These are noted on page 114: they include a proposal to establish a system for the registration, through a single application, of a design which would attract automatic protection within the entire territory of the European Union, as well as a short-lived unregistered design right which would run in addition to similar rights which may be conferred by national law.

5.4.1 Criteria of registrability

Designs can be protected in the United Kingdom by registration under the Registered Designs Act 1949 (RDA). To be capable of registration a design must be applied to an "article", be "new" and have eye appeal.

Design "means features of shape, configuration, pattern or ornament applied to an article by any industrial process...." This may be either a two- or a three-dimensional design.

A design will only be registrable if it is to be applied to a specific article. For this purpose "article" is defined as "any article of manufacture", including "any part of an article if that part is made and sold separately". Each different article on which the design is to be used must be the subject of a separate application for registration.

A design must also be "new" in the United Kingdom at the date of application. It will not be new if the same design has already been registered or made available to the public in respect of the same or another article. Insignificant changes to an old design do not normally create a new one, but one striking feature together with a collection of familiar features will normally suffice.

5.4.2   Exclusions from registration

Under the RDA, any feature of shape or configuration of an article which is dependent on the appearance of another article of which the article is intended by the author of the design to form an "integral part" is excluded from registered design protection. This is known as the "must match" rule.

5.4.3   Duration of registered design

The maximum life of a registered design is 25 years. This is made up of an initial term of five years, after which the registration is renewable for further periods of five years, subject to payment of the renewal fees.

5.4.4   Ownership of registered design right

The author of the design is the first owner of the registered design right, except where (i) a design is created by an employee in the course of employment, in which case the employer is the first, or (ii) where the designer is hired to create a design in return for a fee, in which case the person who commissions the designer is the right's first owner.

5.4.5   Infringement

The owner of a registered design can prevent anyone else from making, using or selling an article which bears that design or one which is virtually the same. As with patents and trade marks there is no need to prove copying.

5.5 Unregistered design right

The unregistered design right (UDR), introduced in 1989, was intended to encourage investment in the creation of designs for industrial products even if they lacked the criteria of registrability. The UDR is like copyright in that it need not be applied for; but, like copyright, it only prevents copying of the design but gives no protection against the use of independently generated designs.

In order for a design right to exist there must be a "design" which is "original" and the design must have been recorded in a "design document". Each of these requirements is examined in detail below. Design is defined as "the design of any aspect of the shape or configuration (whether internal or external) of the whole of part of an article" but does not extend to surface decoration and covers only three-dimensional designs. Unlike registered design-protected products, parts of articles need not be made and sold separately if they are to enjoy UDR protection. Since there is no requirement that the design has "eye appeal", aspects of a design which are purely internal and which are not visible may still enjoy protection.

5.5.1   The need for originality

The design right only subsists in original designs, those which are not copied but are the independent work of the designer. A design will not be "original" if it is "commonplace in the design field in question at the time of its creation".

5.5.2   The design document

A design right cannot exist until either (i) the design has been recorded in a design document or (ii) an article has been made to the design. The design document records the design: it may be a written description, drawing or photograph of the design.

5.5.3   Exclusions from unregistered design protection

The UDR does not subsist in any feature of shape or configuration of an article which enables it to be connected to or placed in, around or against any other article so that either article may perform its function. The reason for this is that subsequent designers should not be constrained in their work by any legal limitation upon the ability of their article to connect to, or fit in, another article. This is known as the "must fit" rule.

In addition, UDR does not subsist in features of shape or configuration of an article which are "dependent upon the appearance of another article of which the article is intended by the designer to form an integral part". This is the "must match" rule and it is similar to that which pertains for registered designs.

5.5.4   Duration

An unregistered design right lasts for either 15 years from the end of the calendar year in which the design was created or 10 years from the end of the calendar year in which articles to the design were made available for sale or hire, whichever is the shorter.

5.5.5   Ownership of UDR

The designer is the initial owner of the UDR, except where (i) the design is created by an employee in the course of employment, when the first owner is the employer, or (ii) the designer is commissioned to create the design in return for remuneration, in which case the commissioner is the first owner of UDR.

5.5.6   Infringement

The owner of the UDR can only prevent others from copying the design but cannot stop the use of a design which has been independently conceived. The infringing copy need not be identical; small changes will not prevent a copy being an infringement.

5.5.7   Licence of right

During the final five years of the period of UDR protection everyone is entitled as of right to a licence to manufacture or sell or hire an article made under that design. If the terms of the licence cannot be agreed between the parties, they will be imposed by the Comptroller of Patents, Designs and Trade Marks.

5.6   European design reform proposals

Two proposals concerning design protection in the European Union were tabled in 1993. Both are now being actively pursued. They relate to a Regulation establishing a Community-wide design right and a Directive to harmonise the different national design protection laws which may be found in the 15 jurisdictions of EU Member States. Since the implementation of these proposed reforms is likely within the near future, this chapter offers a summary of their provisions.

The Directive will not only harmonise the design registration laws of EU Member States but will also align those national laws with the proposed regime for registered Community design rights. The Directive covers registered design rights only, since few Member States have UDR laws. National laws, once harmonised, will co-exist with the Community design right. Since the Directive only harmonises the law relating to registered design rights, national laws relating to unregistered design rights, trade marks, patents, utility models and unfair competition will be unaffected.

The Regulation will create a Community-wide system of protection for designs. Two-tier protection is accorded to industrial designs as follows:

- an unregistered Community design right which comes into being automatically when the design is first made available to the public. This right will last just three years and will protect a design which falls within a one year grace period during which an application for registration may be made, as well as short-lived designs which do not justify the cost of registration.

- a registered Community design right which comes into being when registered following a single application to the Community Design Office in Alicante. The registered design right will last in all 15 member states for a minimum term of 5 years, renewable at 5 year intervals until the expiry of the maximum term of 25 years.

For European purposes, "design" will be defined as the outwardly visible appearance of the whole or part of a product resulting from the specific features of the lines, contours, colours, shape and/or materials of the product itself and/or its ornamentation. The Regulation and Directive thus require protection of the *appearance* of a product: it is the appearance of the whole or a part of the product which must be visible, not the part of the product itself. As under United Kingdom law, a design cannot be protected by a Community design right if the features of its appearance are dictated by its technical function.

The draft Regulation provides that the design of a product which constitutes a part of a complex product will be protected in itself as a design. "Part" in this context means a component part of a complex product. Designs of products constituting parts of a complex product (for example, a piece of machinery) will only be considered to have satisfied the novelty and individuality if (i) when incorporated into the complex product, it remains visible during normal use of the complex product and (ii) the visible features of the design of the part fulfil the usual novelty and individuality requirements. If a part of a complex product is not visible during normal use of the complex product, that part cannot be protected as a Community design in its own right. "Normal use" means use by the end user and will not therefore include maintenance, servicing or repair.

To qualify for registered or unregistered design protection, a design must be "new" and have an "individual character". A design will be new if no identical design has been made available to the public before the reference date and it will have been made available to the public if it has been published, exhibited, used in trade or otherwise disclosed. A design has an individual character if the overall impression it gives the "informed user" differs from the overall impression of another design, while according equal weight to common features and differences. The range of designs with which the design seeking protection will be compared are those designs which have been commercialised.

No Community design right will subsist in a design which must be copied so as to enable the product (i) in which the design is incorporated, or (ii) to which it is applied to be mechanically assembled or connected with another product or mounted in, on or around another product so that both products perform their function. This exception corresponds to the "must fit" exception under UK design law.

Further, no rights conferred by a design right can be exercised against third parties if:

- the product incorporating the design, or to which the design is applied, is part of a complex product upon whose appearance the protected design depends
- the third party uses that design to repair the complex product and restore it to its original appearance
- the public is informed, by means of trade marks, trade names, and so on, of the origin of the product used for repair
- the design rights holder has been told of the intended use of the design
- the design rights holder is offered fair and reasonable remuneration for the use of the design.

In this context, no guidance is provided as to what constitutes a fair and reasonable remuneration.

A one year grace period will be introduced for design applications, during which a disclosure of a design to the public will not defeat a subsequent application to register it on the grounds of lack of novelty. During the grace period the design will attract unregistered design protection.

5.7 Trade marks

5.7.1 Unregistered trade marks

Any word, mark, logo or other distinctive sign which is applied to or associated with the products or services of a trader enjoys the protection of the common law against competitors who use them in order to deceive the public into thinking that there is any connection between their goods or services and those of that trader. This protection is provided through an amorphous body of case law known as the law of "passing off". No registration or other legal formalities are required before an action for passing off is commenced, but it is a difficult and uncertain legal remedy, which is usually expensive to bring. It is therefore recommended that, wherever possible, any mark, sign, name or logo which can be said to attract business or goodwill be registered as a trade mark.

5.7.2 Registered trade marks

Under the Trade Marks Act 1994 a wide range of signs and words, effectively embracing anything which can be graphically recorded, are potentially registrable as trade marks in the UK. Since the law of the UK is now harmonised with that of other European Union countries, it is usually the case that any mark which can be registered in the UK can also be registered elsewhere in the EU, so long as it has not already been registered there and so long as it does not carry connotations in other European languages which, while inoffensive in English, are obscene, deceptive or otherwise objectionable.

5.7.3 General observations concerning trade marks

A registered trade mark runs initially for ten years but it is indefinitely renewable, thus providing much longer potential protection than patents or copyright. The monopoly granted by the trade mark right does not stop rivals manufacturing the same product as that to which the registered mark is applied, but it does stop competitors dressing up their products in the same or a confusingly similar way.

Before seeking to adopt a trade mark, or to register it, it is strongly recommended that a search of existing registered and used marks be conducted. If a mark is selected which is not clear for use, a new product may have to be withdrawn or renamed at a very late stage in order to avoid legal liability for trade mark infringement, perhaps just after it has been launched amid a fanfare of expensive publicity.

5.7.4   The European Community trade mark

From 1 January 1996 the Office for the Harmonisation of the Internal Market (Trade Marks and Designs) has been open to receive applications for a Community trade mark (CTM). The CTM is much like the UK national trade mark, except that (i) it covers the entire territory of the European Union, (ii) it can only be licensed or assigned for the entire EU and (iii) it can only be granted in respect of marks which have not already been registered by other people in any of the national trade mark registries of the EU.

The Office (usually referred to as the Community Trade Mark Office or just plain CTMO) is located in Alicante, Spain. Applications can however be channelled through any of the national trade mark offices of the EU or through the Benelux Trade Marks Office.

5.7.5   International trade mark protection

It is not yet possible to apply, through a single application, for trade mark protection throughout the world. There are however two international filing arrangements which should be noted.

The Madrid Agreement Concerning the International Registration of Marks of 1891 enables a person who has already obtained a trade mark registration in one Agreement country to extend it to others through a single application to the World Intellectual Property Organization's International Bureau in

Geneva, the application going through by default unless any of the designated trade mark registries of member states object to the application within a limited period of 12 months. The Madrid Agreement is of only limited utility, since no English-speaking country in the world, no country in the length and breadth of North and South America, no country in Scandinavia has signed the treaty – not to mention Japan.

The Madrid Protocol of 1989 is intended to cure the defects of the 1891 Agreement, by enabling among other things an international application to be pursued without the need to have a granted trade mark in the first place. At the time of writing, however, only a few countries have acceded to it. If an application under the Madrid Protocol is rejected, it can be divided up into separate national applications which enjoy the priority date of the original Protocol application.

Lists of signatories to the Madrid Agreement 1891, and to the Madrid Protocol 1989, appear in Appendix 7.

5.8 Sources of professional advice

- The **Institute of Trade Mark Agents (ITMA)** is the professional body for trade mark agents in the UK. It publishes a geographical list of members, and produces a short guide for those considering registering a trade mark.
- The entries at the end of Chapter 4 are equally relevant here. The **Patent Office** and the members of the **Chartered Institute of Patent Agents** give advice on all kinds of intellectual property rights.
- The government-supported **business information centres** often offer professional advice on design matters. Contact numbers for these centres are given in Section 8.7.7.

Contact details of the organisations mentioned above are given in Appendix 1.

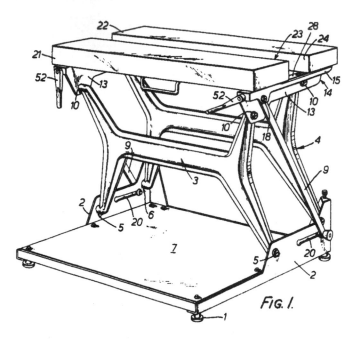

Figure 8. British patent number 1267032. One of Ron Hickman's patents for the Workmate portable workbench, applied for in 1968.

6
Developing an Invention through Prototyping and Design

Charles Dawes and Richard Paine

This chapter covers the "nuts and bolts" of turning an invention from a patented concept into a product ready to be launched on the market. Many inventors without the backing of professional colleagues or any experience of business, on reaching this stage, simply do not know what to do next – this chapter is for them! Every step forward seems to raise new questions, and some of those which are most often asked are dealt with in the sections below:

- will my invention sell?
- do I need a prototype?
- can I sell my invention as a concept only?
- does a prototype need to be of high quality?
- how can I get a prototype made?
- how do I get drawings made of my invention?
- do I need to work on the visual appeal of my invention?
- how much do I need to know about the development process?
- how do I get my invention into manufacture and onto the market?
- how do I find a company to develop my invention?

By the time that issues such as prototyping, design and manufacture are being considered a considerable amount of work should have been carried out on the invention, including:

*   establishing that the concept is unique
*   research to ensure that the invention has an end market
*   developing the original idea in the light of the research
*   protecting the invention by a patent application or other forms of intellectual property rights.

Advice on these steps is given in other chapters, and pointers to sources of information in Chapters 7 and 8. It is now time to move from the ideas and protection stages of the invention into a new phase. This next stage is probably the most difficult of all, because it requires the inventor to sell the invention to other people who may have different views, who have not been involved in the project from an early stage and, perhaps even worse from the individual inventor's point of view, who already have many other ideas to look at, including their own.

There is one paramount rule: adopt a totally professional and business-like approach. Others will be looking at the invention as a cold business proposition, so it is essential that the inventor does the same. In particular, do not disclose any details of the invention except in confidence, after the person or company to whom you disclose it has signed a confidentiality agreement.

6.1 Will my invention sell?

There is no easy answer to this. For some apparently excellent ideas, the time is simply not right. Perhaps the circumstances which generated the original idea have changed, perhaps the market gap has closed or has been filled by another product. It is therefore vital at this stage to review all the work carried out to date, especially to see if the market has changed. Do not carry out the same research in the same way. This will simply give the same answers. Instead, ask searching questions and answer them fully and honestly. Help with this is given in Chapter 2. Having done this you can then ask the question – is it worth proceeding?

Other inventions immediately "fall on their feet". It is not the quality of the idea, but how it fits into the marketplace, which determines success. An important step towards an invention "falling on its feet" is the development of an appropriate prototype.

6.2 Do I need a prototype?

A prototype is a very useful way of demonstrating that the invention actually works. Inventors should not be deterred from proceeding with an idea simply because they do not have a prototype, but it is often easier to explain the thinking behind an idea with a working prototype or model on which to focus. A prototype is not essential to begin with and it is quite possible to persuade others of the value of an invention without one.

The need for a prototype depends on the complexity of the concept. One should ask whether an outsider, hearing about it for the first time, would have difficulty immediately understanding the advantages of the concept. If so, then careful consideration needs to be given to the various methods which can be used to demonstrate the invention. The main reasons for deciding against producing a prototype, and the appropriate actions to be taken in that event, are as follows:

- the invention is too large and/or too expensive to manufacture at this stage – in which case one should consider the possibility of producing a scale model or even a "virtual" computer model
- the concept is simple and easily explained – ensure that all written material, including drawings, are clear and well presented
- the inventor is unsure of the precise physical attributes of the idea and further thought needs to be given to at least one way in which the concept can be realised. For example, it is not necessary to produce prototypes in a range of different materials, colours or sizes. One simple prototype which demonstrates the main idea is adequate; the rest can be explained briefly in writing and by drawings (produced with

expert help if necessary). If an idea cannot be visualised in a physical shape, then serious consideration needs to be given to whether it has a real use.

6.3 Can I "sell" my invention as a concept only?

Yes, but it is important that the written presentation fully communicates the benefits of the idea. The patent abstract is a starting point, but that for which the patent is claimed is only a legal description of the invention. In this context it is vital to ask of the invention the following questions:

- what is it?
- what does it do?
- how does it do it?
- why does it do it?
- what are the advantages to the manufacturer?
- what are the advantages to the end user?

These are all points which need to be communicated to anyone who may be interested in turning the concept into reality. Many inventors try to explain their concepts without establishing these basics – to the bemusement of their listeners and to their own disadvantage.

If the idea is to be sold as a "concept only" there must be a concise written description of the concept in clear language. This should cover the basic points, with a more detailed explanation to follow if necessary, and a short summary. Diagrams and drawings are also important. They should be neat, accurately labelled and with an explanatory title. The product description, drawings and diagrams do not have to be professionally produced, but attention should be paid to the presentation. Ideas jotted down on the back of envelopes are unlikely to impress busy company executives. Attention to detail demonstrates that the inventor has put considerable thought into the concept; this is important to generate confidence in a potential buyer.

6.4 Does a prototype need to be of high quality?

No, the quality of the prototype does not matter, providing it is in working order and demonstrates clearly the main purpose of the invention. Further explanation of the concept may be needed, in which case the guidelines on written presentation given above should be followed.

6.5 How can I get a prototype made?

There are companies which offer professional "modelmaking" services. Although they tend to be expensive, it could be worthwhile contacting local ones to check prices and their areas of interest and expertise – check the entries under "Production engineers" and "Models – architectural and engineering" in the *Yellow Pages* directory.

Do not forget that personal contacts can be useful. Ask friends and relatives if they know of anyone with the practical skills required. There may be someone who has a relevant hobby, or can adapt skills learnt at work.

Although it will be necessary to give a broad idea of the invention to a potential prototype maker, do not transmit any confidential information until the individual has signed an undertaking neither to reveal the details of the invention to a third party nor to use them on their own behalf. This is important even if the prototype maker is a friend, if only because it impresses upon him or her that the invention is a serious proposition. A sample disclosure undertaking is given in Appendix 5.

If a friend, relative or acquaintance has agreed to make the prototype, ensure that this is not to be done at work or on another business' premises, as this could lead to a potential dispute of ownership, and make sure it looks sufficiently professional for the purpose to which it will be put.

If personal and local contacts do not seem likely to produce a satisfactory prototype which will sell your idea, and the cost of modelmaking services is unaffordable, there are other possibilities which you can explore. An increasing number of universities are using their laboratories and workshops, and the expertise of their staff, to offer a service to inventors who need skilled help with developing an idea. As examples of university-based product development organisations there are Brunel Innovation, and Product Development Tayside, with programmes to help both individuals and small businesses. Product Development Tayside is only one example of the many recent ventures supported by local groups interested in promoting innovative enterprise, in this case a university, a community business organisation and the Regional Industrial Office. These kinds of groups may provide help with funding, and usually try to keep costs low, but after the initial assessment of your invention's potential you should expect to pay for their services. The business information services described in Chapter 8 can help you find institutions to approach for further details.

Other possible sources of help with prototyping are the regional networks such as NIMTECH and WEMTECH. Your local business information centre can give you contact details for these and any other organisation offering this kind of help to innovators.

6.6 How do I get good drawings made of my invention?

A patent agent will usually produce drawings of the invention which are sufficient to explain it for the patent application. If it is necessary to produce more interesting or artistic drawings, again, look around among friends and family. Someone with a good level of drawing skills may be able to help.

Yellow Pages directories list freelance designers (under "Design consultants") who can usually produce drawings for an agreed fee – ask to see their portfolio to check that their style matches your requirements. For engineering drawings, an engineer in the relevant field may have the necessary skills. The product development organisations mentioned in the section above will also be able to help with drawings.

6.7 Do I need to work on the visual appeal of my invention?

There is a common misconception that the look of a product is in some way separate from its function. This is not usually true. The vital aspect of a product is its function – what it does and how – and the appearance grows from this. In most cases it is not necessary for an inventor to commission design work to make the invention more attractive. It is also not usually necessary to spend a lot of time and effort on packaging design, or contriving a logo for the invention, unless this is an integral part of the idea.

Even if spending time and money on the look of your invention is not strictly necessary, you may still be interested in this aspect. If so, help is available from the kinds of product development organisations already mentioned in Section 6.5, or from some of the business information centres listed in Section 8.7.7 – the Business Links, for example, usually have a Design Adviser. The protection of your rights in these kinds of creative products is dealt with in Chapter 5.

6.8 How much do I need to know about the development process?

How much information do I need to have on the potential for my idea? For example, is it necessary to know production costs, retail costs and details of materials?

It is unlikely that the inventor will have all this information to begin with, but facts and figures are always useful so it is sensible to gather as much information as possible, especially on similar or competitive products which are already on the market. For example, if all products with a similar function are manufactured in plastic there will be good reasons for this in terms of cost, function and consumer demand. The inventor needs to investigate the dynamics of manufacturing and marketing the invention, in order to avoid proposing a material which is impractical or a manufacturing technique which is too expensive for the market. The World Metal Index is one example of a source of data on specific materials, and the next chapter gives guidance on finding technical information like this. Relevant trade associations can often give brief factual information free of charge, or point to another source of help.

Good records of all the information collected should be kept. It is worthwhile consulting a professional if the figures are complex, and Chapter 2 gives advice on where to go for help. Remember, however, that companies vary in the way in which they cost the various elements of their businesses. If a manufacturer is interested in an invention they will produce their own figures based on their own new product development procedures.

6.9 How do I get my invention into manufacture and on to the market?

There are many different routes to manufacture of a new product, and the options open to the inventor vary with individual circumstances. The choices available for starting or expanding a business are looked at from a financial viewpoint in Chapter 3. Some legal aspects are discussed in Chapter 5, and detailed help with preparing a business plan and interesting others in the product is provided in Chapters 2 and 3.

6.10 Finding a company to exploit the invention

Section 2.3 of the marketing chapter suggests criteria for selecting companies which might be approached to exploit the invention. This section suggests how to find a suitable company for yourself.

The companies which are most likely to take on inventions are those seeking new product ideas for development within a certain market. The names of potentially interested companies may be found by searching through relevant trade publications and company directories, and classified telephone directories. Both details of what publications to use, and where they can be found, are given in Chapter 8, which deals with the sources of business information.

Small to medium-sized companies probably offer the best prospects for the individual inventor. Large companies and market leaders may appear to be worthwhile, but in reality they are very rarely willing to deal with individual inventors. They have their own design and development departments which are perfectly placed to meet their development and market plans. Unsolicited ideas, of which they receive a large number, are very unlikely to fit into these plans. The cost of introducing an outside idea will nearly always be greater than the gains, due to their financial commitment to existing manufacturing and marketing. In short, they nearly always reject ideas from individual inventors, often claiming to have, or be developing, a similar product.

Start with the local area and select a few small or medium-sized companies. They may not be household names, but a new invention may be what they need to make their business grow! Do not approach companies which already have a product on the market which yours would compete with. Look for those which are in the same or a similar market area, and could take on the invention as a "line extension" or for diversification.

If the invention is complicated or expensive to manufacture, or requires specific manufacturing techniques, then try to find companies which would be capable of fulfilling the manufacturing requirements, without having a product similar enough to compete directly with yours. If the manufacturing is fairly straightforward, but distribution and marketing requires specific skills, then marketing-led companies are more appropriate. Again these can be identified with the help of the information provided in Chapter 8.

From your research a short list of potential contacts should be produced including the name and title of the appropriate manager/director, the full company name and correct address. Directories go out of date very quickly and management changes. Ring the company and check the address and contact name with the switchboard.

The initial approach to the company should be a short letter to find out whether the company has any interest in looking at new product submissions from outside sources. Some companies have a strict "no external submissions" policy, others require a particular procedure to be followed, some will only look at submissions from their agents or other companies. Advice on compiling a profile of the product is given in Chapter 2. No confidential details should be revealed at this stage, but it will be necessary to give some information about the invention so the company can decide whether they wish to proceed to the next stage of a confidentiality agreement. Include a contact name, address and telephone/fax number.

If no reply is received from the company, a short telephone call or a second letter to the same person should be sent, reminding them of the first letter and asking politely for an answer. Remember, companies often take some time to consider outside ideas, other people within the company may have to be consulted, they may be on holiday, or be very busy. Ideas received from outside inventors do not take priority over their day to day work. Give them time to consider the invention and reply, then telephone or send a gentle reminder. If telephoning, be prepared to speak to the manager's secretary, they are usually very helpful, will probably have seen the initial letter and be handling the reply. Frequent telephone calls asking whether the letter has been received, when a reply can be expected, whether they are going to accept the idea and so on, are likely to irritate busy companies and should be avoided.

Most inventors receive a lot of rejections. Carefully read each letter, and use any constructive criticism to improve the invention. Do not take too much notice of effusive praise; if they turn the invention down they're not interested – they're just trying to be kind. Research some more companies, moving outside the local area if necessary, and start the process again. When you feel you have exhausted your ideas, or earlier if you are ready to involve professional help, you could approach some of the business information services mentioned in Chapter 8. They will put you in touch with an adviser who can review the ground you have covered, and then suggest other contacts. There may be a charge for this, depending on the work involved.

When you find a company which is interested, arrange a meeting at which more information about the invention can be given, the prototype demonstrated and so on. Confirm the time, date and location in a short, business-like letter. You may not be allocated more than half an hour, so ensure that the presentation is brief, interesting, informative, and relevant. Again, listen to any constructive criticism and learn from it. If the company turns the idea down, continue to try elsewhere, building on what has been learnt. There is advice in Chapter 2 on preparing documentation for the presentation.

If a company is genuinely interested, further meetings will have to be arranged. Remember, maintain a professional attitude at all times. Confirm important points resulting from meetings and phone calls in writing, stay in touch, stay calm! The next stage is the minefield of negotiations. You will find more advice on this and related aspects of developing your invention in Chapter 2.

6.11 Advice for inventors

- The **Institute of Patentees and Inventors** is a non-profit making association which offers its members advice and guidance on all aspects of inventing. Members also have reduced-cost access to information and exploitation services.

Contact details of all the organisations mentioned in this chapter are given in Appendix 1.

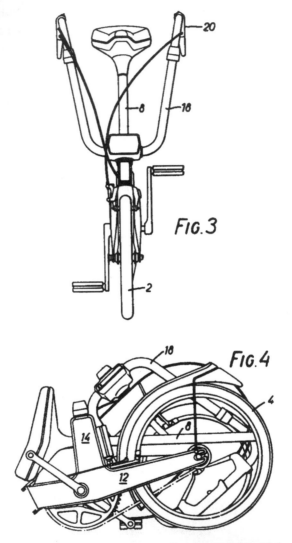

Figure 9. British patent number 1460565. Harry
Bickerton's foldable bicycle patent, applied for in 1972.

7
Sources of Technical Information

Information Officers at the British Library

This chapter describes types of scientific and technical data which the inventor may need and shows how this information can be located. Even those who feel they have all the information they need should use the checklists to make sure they have not overlooked anything.

The chapter is in six sections:

- information needs
- types of information
- finding information
- choosing the best access route
- guide to libraries and other information services
- online information.

7.1   Information needs

Inventors are more likely to need technical data than information on basic scientific principles, though, where the impossible seems to have been achieved, such as a perpetual motion device, it could save time and money to look at the basic equations before attempting to take the idea further.

If the development of an idea depends, say, on finding a material with certain characteristics, the need for information is self-evident. There may be a need for other data which is not so obvious but equally important at the technical development stage, so the inventor should be asking questions such as:

*   will my idea work?
    -   can I quantify the benefits of my idea as compared to those benefits which can be derived from existing products or systems?
    -   do I know enough about the properties of the material I propose using?
    -   do I know the sources and the availability of relevant materials/components?
    -   do I know of all the standards or regulations involved (e.g. on reliability or safety aspects)?
    -   are there any proposed changes in legislation relevant to me?
    -   can I get useful pointers to possible problems in operation from work in similar areas?

Checklists of things to think about during the development process appear in the chapters dealing with marketing, development and finance.

The most important question for the inventor to ask, as soon as the basic idea is formulated, but before time and money is spent on investigating details, is:

*   has anyone thought of my idea before?

In practice, research on the novelty aspects of the idea and working out of design details often interact and then continue in parallel. A patent may be found covering the process in the form first thought of, but the inventor may then think of a development of it which is new and patentable. Conversely, research on a material's properties may reveal new possibilities which the inventor then wants to incorporate in the design, but any additional features will necessitate a further check for patent infringement. Fortunately it is easy to go back and forth from the patent records to other scientific and technical information, as patents are found in the major technical libraries.

Patents have a unique dual role: they both document intellectual property rights and make public a vast amount of detailed technical information and innovative ideas. They are included as an information source in this chapter both for the essential novelty check at an early stage in the inventive process and as a valuable category of technical literature in their own right.

7.2 Types of information

What information is available? The varying forms and places in which technical information is found are in part a result of the way knowledge is disseminated. Table 1 shows how knowledge starting as an idea in someone's head is developed and packaged until it may eventually appear as data in a standard reference work.

Information from any part of the process can be located via directories or indexes in libraries. The information may have been stored in a variety of communication media: a journal article may be available online or on CD-ROM, as well as in print in the original journal, or photocopied from it.

Once information needs have been identified, the next step is to decide on the most appropriate source to go to, and how to access it.

7.3 Finding information

Information can be found by:

- using material to hand at work or at home (e.g. back issues of periodicals)
- using online information at work or at home (see the sections in this chapter on the Internet and online databases)
- using personal contacts
- using facilities offered by organisations to members

Table 1. Sources of information

Sources of information	Guides to sources
Research, invention, development	Directories of research in progress; directories of research centres and organisations
Discussion in person or at a distance/worldwide via the Internet	Personal contacts; professional organisations; guides to Internet discussion lists
Conferences (formal presentations and informal talk)	Directories of forthcoming conferences, and reports on conferences held; professional organisations; diary columns in appropriate periodicals; databases
Reports and theses – often unpublished, may be available only from the organisation which produces them. Sometimes confidential, for internal use only	Indexes to reports and theses. Indexes to "grey literature"
Patents – full descriptions of inventions	See Chapter 4 for a full discussion
Journal articles – may be formal articles published in refereed learned journals, or notes and news in a trade magazine	Abstracts and indexes. Browsing through appropriate journals, using any annual indexes they produce
Review articles – critical evaluations and compilations of previous work on a topic	Index to scientific reviews
Books – popular, introductory, practical or learned treatises	Subject and general bibliographies; catalogues of appropriate libraries, including remote libraries accessible online
Handbooks	Bibliographies; guides to reference books
Standards	Catalogues of national and international standardisation bodies
Encyclopaedias	Guides to reference books; library catalogues
Trade literature – product catalogues, user manuals, instructions and technical data sheets	Directories of manufacturers and suppliers; guides to trade literature; advertisements

- visiting local libraries or other information centres (see the next section for more detail)
- ringing Science Line (see Appendix 1 for more information)
- contacting information centres by telephone, fax, e-mail or letters. They will locate the information, point to other possible access points or do actual research (there is often a fee for lengthy enquiries).

7.4 Choosing the best access route

There is often more than one way to find the information you require. Time and money can be saved by choosing the most appropriate source for any particular need. Each type of source is good for some needs but less suitable for others. The most direct route is not always the best for you, as the examples in Table 2 demonstrate.

7.5 Guide to libraries and other information services

The UK is fortunate in having a long-established information network freely available to all and well-organised for retrieval of specific pieces of data. The next section will describe the library and information services which are open to the general public or which are at least partly accessible to them:

- British Library Science Reference and Information Service
- British Library Document Supply Centre
- public libraries
- the Patents Information Network
- England
- Scotland
- Wales
- Northern Ireland
- co-operative information services
- universities

Table 2. Strengths and weaknesses of some information sources

Examples of information sources	Strengths	Weaknesses
Standard handbooks	Reliable, easily accessible data	Figures may need to be interpreted
Trade catalogues	Current prices and specifications of goods in the marketplace	Need good selection to cover range of products; comprehensive collections are few and far between
Online search bureau	Fast, comprehensive, expert search skills	Relatively high cost
Manual searching at a technical library	Free; allows chain of references to be followed	May not be available locally; labour-intensive; probably not comprehensive coverage
Personal contact	Free access to information and advice	May not have fully up-to-date and comprehensive information
Science Line	Access to experts in every branch of science for the cost of a local phone call	Designed for the general public; may not be able to give enough detailed information
Online information via a personal computer connection	Potentially fast, with wide coverage	Can be time-consuming, expensive and unsuccessful for a beginner

- government information services
- independent organisations.

7.5.1 The Science Reference and Information Service (SRIS)

The British Library Science Reference and Information Service is the national collection for science and technology, including patents. Much material, mostly recent, is on open access. The collection is for reference only and materials cannot be borrowed. There are no restrictions on entry. It is housed in central London, but many enquiries can be answered by phone or letter. A detailed description of its resources follows, both because of its importance as the major centre of innovation research in the country, and also because its comprehensive collections demonstrate the wide variety of published information, much of it available in other large libraries. Scotland and Wales have their own national libraries, which are described in the sections on pages 153-4 about facilities in those countries.

How to use SRIS

SRIS can provide help for personal callers and answer remote enquiries by letter, telephone, fax or e-mail. Using SRIS for business information is covered in Chapter 8. Personal callers will receive any help they need to use the material in the library, such as:

- identifying a reference in a book or journal
- using indexes and abstracts
- carrying out a subject search
- using the CD-ROM databases.

Staff at the enquiry desk can give brief answers to questions on scientific and technical points, but users are expected to carry out their own research.

On the patents side staff will explain how to carry out a search to see if an invention is new; this will include guidance if necessary on the patent classifications, the layout of a patent specification and the most appropriate CD-ROM or microfiche databases to use. Similar help is given on using the patent material as a source of technical information. Examples of patents information provided free to both personal callers and those at a distance include:

- finding out if a British patent is still in force
- finding an English language equivalent of a foreign language patent
- searching the Designs Index (1962 to date) by applicant's name.

All such help is free; enquiries about intellectual property topics may result in the delivery of a printed list of relevant references, given free to enquirers. When a longer patents search is wanted, and the enquirer is unable to do it, further help may be offered on a priced basis.

The range of priced services is described briefly under SRIS services below.

Location of stock
Holborn
The Holborn Reading Room is located at 25 Southampton Buildings, London WC2A 1AW, close to Chancery Lane underground station: tel. 0171 412 7919, fax 0171 412 7480, e-mail patents-information@bl.uk for patents enquiries or tel. 0171 412 7494, fax 0171 412 7495, e-mail sris-centre-desk @bl.uk for science and technology enquiries. The reading room is open 9.30-21.00, Monday to Friday, 10.00-13.00, Saturday.

The Holborn Reading Room holds those patent specifications which were published in the course of applications to seek to protect an invention in Britain plus related gazettes, indexes and other materials. The collection comprises the series of British, European Patent Convention and Patent Cooperation Treaty specifications. Holborn also holds books and journals on engineering and physical sciences and technologies, as well as intellectual property. There is a business collection which holds information on companies, markets and products; this part of the collection is described in detail in Chapter 8.

Chancery House
Chancery House (across the road from the Holborn Reading Room) is an annex holding many foreign specifications: tel. 0171 412 7902, e-mail patents-information@bl.uk for details. It is open 9.30-17.30, Monday to Friday.

Older material, and that for the smaller countries, may be held either in vaults at Holborn or at a further store at Micawber Street.

Aldwych
The Aldwych Reading Room, 9 Kean Street, London WC2B 4AT, tel. 0171 412 7288, fax 0171 412 7217, e-mail sris-aldwych-desk@bl.uk holds life science material and technologies, including biotechnology, medicine, agriculture, mathematics, astronomy and earth sciences. It is open 9.30-17.30, Monday to Friday.

Principal searching tools
SRIS has an online public access catalogue (OPAC) which can be consulted either at Holborn or Aldwych or remotely from some UK libraries.

The OPAC allows keyword or author access to, generally speaking, the following:

- all journals, including those discontinued
- all books published from 1969 onwards
- all dictionaries
- many books concerning intellectual property which were published before 1969.

All the material found on OPAC (except for series of patents, or official intellectual property gazettes) will have an SRIS classification mark given to them. This unique classification is used to classify material on the shelves in three sequences, (B) books, (P) periodicals or journals, and (A) abstracting journals or bibliographies. Everything given a particular (B) location will be shelved in no particular order at that classmark, while there are specific locations for each (P) or (A) periodical, abstracting journal or bibliography. There is a detailed *Subject index* which can be used to identify relevant classmarks.

Older material can be found in a 1931-69 card catalogue to which there is author and subject access. There are bound catalogues for the 1855-1930 material.

There are numerous CD-ROM databases which can be used to identify material of interest, much of which would be held at SRIS. They are usually on annual discs and only cover the last few years. They will be mentioned in the relevant sections below.

There are also numerous abstracting journals, many of which have not been superseded by CD-ROM, or which cover years not available on CD-ROM.

There are some microfiche indexes, especially for patents.

Categories of material
Conferences
Many published proceedings of conferences are held at SRIS, as books, periodicals or within periodicals. They are indexed on the *Boston Spa conferences* CD-ROM, which gives reference to the source. Any journal locations at SRIS are included in *Boston Spa serials*. For those wishing to attend conferences in person, there is a listing of forthcoming conferences on scientific, technical and medical topics, *World meetings*. It is published in two volumes, one covering the United States and Canada, and one the rest of the world. Revised and updated quarterly, it has a keyword index of topics.

Exhibitions and conferences as such are listed in various publications held at SRIS classmark (P) AA 39.

Articles
SRIS holds about 28,000 periodicals. Relevant periodical articles can be identified by using abstracting journals, both on paper and CD-ROM, mentioned below.

Reports
SRIS holds many reports series issued in Britain, but not foreign series. Many of the latter can be identified and ordered from the Document Supply Centre, part of the British Library, through a local library. Many reports and other forms of "grey literature" (not produced by traditional publishing channels) such as

translations can be found by using the *SIGLE* CD-ROM which indexes material from 1980.

Abstracting and indexing journals

The abstracting and indexing journals are arranged by subject on the open shelves. Some of these are duplicated and made more accessible by annual CD-ROM products. The CD-ROM products include *Inside Science PLUS*, which indexes article titles from 13,000 journals from 1993; *Science citation index*, from 1986, which indexes keywords in numerous journal articles as well as the citations made in them; *Compendex*, from 1989, on engineering and *Inspec*, from 1989, on electrical engineering; and *Chemical abstracts*, from 1987.

Reviews

There are numerous books or articles which discuss the more important work done in a particular narrow field. These are indexed in *Index to scientific reviews* from 1986.

Research in progress and theses

Many research projects being carried out by British academic or postgraduate staff are indexed in the annual *CRIB, Current research in Britain*, which has separate biological sciences and physical sciences volumes. Other kinds of research may be identified by consulting abstracting or indexing tools, or using patents to identify preliminary work already published.

British theses are covered in Aslib's annual *Index to theses accepted for higher degrees in the universities of Great Britain and Ireland*, which has detailed abstracts. The actual theses are often available from Boston Spa. Foreign (mainly American) theses are covered in the monthly *Dissertation abstracts international*, which is available online as well as on paper.

Books

SRIS acquires about 8,000 books annually. There is a strong emphasis on technical books, including such practical subjects as food processing and packaging. Coverage of British output is nearly complete while foreign output, particularly in foreign languages, is more selectively acquired.

Handbooks, dictionaries and standards
SRIS holds numerous handbooks. A good way to identify key works in a field is to use *Walford's guide to reference material*, where vol. 1 covers science and technology. SRIS also holds thousands of dictionaries, arranged by language (if monolingual or bilingual) or by subject (if multilingual).

British and American Society for Testing and Materials (ASTM) standards are held. British standards are listed and indexed in the annual *BSI standards catalogue*, which indicates how close to an International Standards Organisation (ISO) standard the British standard is. A good source for finding relevant standards is the *Standards and EC infodisk* CD-ROM, which indexes standards from many national and international standards organisations. The British Standards Institution is the national repository for foreign standards.

Business materials
The Business Information Service at SRIS holds a major collection of directories and market research reports. These can be used to identify product information or to find data about the major companies in a particular field of manufacturing. More details of this collection can be found in Chapter 8.

Patents
SRIS holds over 38 million patent specifications from 36 patent authorities. Patents are invaluable in searching prior art to determine the novelty of an invention. They are also of interest to those seeking fresh ideas and perspectives. A patent document should have sufficient detail in text and drawings for someone skilled in that branch of industry to be able to reconstruct the invention from the specification alone; it is often the only description of the invention published, or at least the most detailed one. This means that patents are a valuable source of technical information in general, not merely for checking the novelty of an idea. In addition to the very detailed description and diagrams, all

patent specifications must include citations to relevant documents: some include extensive state-of-the-art reviews and discussion of problems encountered. Online searches can use citations of earlier patents to trace the development of a technology and can also extract useful business information from the obligatory statement of ownership (other aspects of online searching are discussed later).

Figure 5 shows the information published on the front page of a patent application document.

SRIS arranges its patents collection by country, publication stage, and then number. Most patent documents are on paper but are increasingly available on microfiche, microfilm or CD-ROM. Shelf guides should be consulted for the exact location as stock moves periodically occur. Those patents applicable to protection in Britain are stored at 25 Southampton Buildings while the more recent foreign patents are stored at Chancery House. These applicable systems are British national patents, the European Patent Convention (EPC) and the Patent Cooperation Treaty (PCT).

Patent authorities for which specifications are currently being received are listed below. Other countries may be held: details can be obtained from SRIS.

ARIPO (regional system for English-speaking Africa)	1985 to date
Australia	1904 to date
Austria	1899 to date
Belgium	1949 to date
Brazil	1922 to date
Bulgaria	1953 to date
Canada	1942 to date
China	1985 to date
Czech Republic (& former Czechoslovakia)	1919 to date
Denmark	1895 to date
European Patent Convention (regional system)	1978 to date
Finland	1942 to date
France	1902 to date
Germany	1877 to date
Hungary	1895 to date

India	1912 to date
Ireland	1928 to date
Japan	1889 to date
Korea	1984 to date
Luxembourg	1991 to date
The Netherlands	1913 to date
New Zealand	1965 to date
Norway	1892 to date
OAPI (regional system for French-speaking Africa)	1966 to date
Patent Cooperation Treaty (world system)	1978 to date
Poland	1925 to date
Romania	1957 to date
Russia (& former USSR)	1863 to date
Slovenia	1992 to date
Spain	1970 to date
Sweden	1885 to date
Switzerland	1888 to date
United Kingdom	1617 to date
United States	1791 to date
Vietnam	1984 to date
Yugoslavia	1984 to date

As the collections are not arranged by subject it is necessary to carry out a search using indexes or databases before looking at the patent specifications on the shelves. Generally speaking, each patent authority can be searched separately. The Inpadoc microfiches allow searching of over 50 patent authorities from 1978 or earlier by patent classification, applicant, inventor or patent specification number. The classification used is the International Patent Classification.

The United States patents issued from 1969 can be searched by keyword on CD-ROM (*CAPS* and *Cassis*), while *Bulletin* is used for searching the European Patent Convention specifications, and *Access* for both Convention and Patent Cooperation Treaty specifications. Online searching generally offers the best searching possibilities and is offered as a priced service. For more information and advice on online searching see Section 7.6.

Specifications are normally in the language of the issuing country, but an English abstract is frequently available; Derwent's microfiche *Patent family index* can be used to find equivalent patents in the English language when these exist. If a full translation is required, Transcript offers a specialist priced patent translation service, and details are given below under "SRIS specialist services".

There is a large collection of reference books covering such points as patent law, inventions and classification schemes.

Like all member libraries of the Patents Information Network, SRIS gives free access to the Patent Office's *Patent Training Package*, a self-help interactive learning system based on laser discs.

Government and official publications

Although SRIS is not the first source for most official publications, it does hold some of those relevant to technology, for example a selection of regulations. While the coverage of regulations is by no means complete, it has important works in areas like machinery, electrical wiring, thermal insulation and health and safety. One specialist type of regulation is that which relates to standards: many standards must be complied with by law where relevant (e.g. electrical plugs). Most relevant areas are indexed in the SRIS classification scheme.

SRIS specialist services

Apart from the collections and the help that is available in the SRIS reading rooms, SRIS has a number of specialist services designed to help those people who want to commission information officers to carry out their research for them. A charge is made for these services and for related support services.

STMsearch

This is a priced service offering scientific, technical and medical information from a team of specialist information officers using over 400 online databases. In most cases the result of the search will be a list of references to relevant articles with summaries. Upon request the staff will also carry out manual subject searches of the literature.

All technologies are covered; topics researched for users recently include metal fatigue, adhesives, miniature pumps, centrifugal compressors, charge coupled devices, and fire resistance of glass fibre. Enquiries can be made by tel. 0171 412 7477, fax 0171 412 7954 or e-mail stm-search@bl.uk.

Health Care Information Service (HCIS)
The Health Care Information Service acts as an access point to a broad subject area, offering customers information on such topics as: clinical and non-clinical medicine, nursing, health service management, allied health disciplines and complementary medicine. HCIS offers a range of products and services, including both printed and electronic publications, search and current awareness services and training courses. Brief enquiries are handled free of charge by the Aldwych Reading Room, which contains the majority of our reference collection on medicine, whilst longer enquiries requiring online database access are handled by HCIS as part of the STMsearch service (see above). Aldwych Reading Room: tel. 0171 412 7288, fax 0171 412 7217, e-mail sris-aldwych-desk@bl.uk, HCIS enquiries: tel. 0171 412 7489, fax 0171 412 7217, e-mail hcis@bl.uk.

Environmental Information Service (EIS)
Set up primarily to serve the business sector, the Environmental Information Service remit has expanded to serve all sections of the community, and is an access point for the whole range of material relevant to environment topics. Brief enquiries are answered free of charge, and detailed research, including the use of online databases, is carried out on a priced basis. EIS tel. 0171 412 7955, fax 0171 412 7954, e-mail eis@bl.uk.

Patents Online
Patents Online has access to databases covering the activity of all major patenting authorities across the whole range of patentable technology. The Patents Online staff combine database searching skills with an understanding of the documentation produced by the patent system and can use these databases to:

- find out what patent applications have already been filed in a particular technical area
- provide a state of the art survey of a technical field by identifying key patents
- find out if the solution to a technical problem is already described in a published patent
- monitor new patent applications filed in a particular area of technology
- monitor new patent applications filed by particular companies or individual inventors.

Patents Online: tel. 0171 412 7903, fax 0171 412 7480, e-mail patents-online@bl.uk.

Patent Express
A copying service geared to the needs of patent users, Patent Express supplies copies of patent documents within 24 hours (3 hours from the Urgent Service). Patent Express also deals with requests for copies of material from SRIS's general stock, or for patent file histories. Patent Express: tel. 0171 412 7918 and 7929, fax 0171 412 7930, e-mail patent-express@bl.uk.

Transcript
A service offering competitively priced English-language translations of any patent specification in the British Library. Other technical documents can also be translated. Transcript: tel. 0171 412 7926, fax 0171 412 7930.

Business Information Service (BIS)
The range of business information available from BIS is described in Section 8.7.1 and Sections 8.1–8.6 describe the different types of published business information in detail.

7.5.2   British Library Document Supply Centre (DSC)

There is another part of the British Library which offers services of interest to those seeking scientific and technical information: its Document Supply Centre, the world's largest source of published information in every field of scientific and technical knowledge, which specialises in providing otherwise hard-to-find items. It has extensive holdings of books and journals and outstanding coverage of conference proceedings, theses and reports. The reports include technical memoranda and research notes on progress, many not published commercially.

DSC encourages use of its vast collections by publishing guides which provide a subject approach to the contents of the various forms of literature. Its titles include:

- *British reports, translations and theses*: published monthly in hard copy, contains sections on mechanical, industrial, civil and marine engineering; it is the only publication to provide comprehensive lists of non-commercially published material from government, industry, universities and learned institutions
- *Inside conferences*: published quarterly on CD-ROM, lists the contents of the published proceedings of around 15,000 conferences each year, which often include the first announcements of innovative ideas and new research findings.

These guides to the world's technical literature are available in many large libraries. Items of interest which are identified through them can then be obtained through the document delivery service or from other sources.

For researchers who need to keep up with developments in their field DSC produces *Inside Science PLUS*, a CD-ROM which offers many features to make retrieval of information from journals quick and easy. It includes 1.3 million articles each year from 13,000 scientific journals, information on local holdings, searchable abstracts, and automatic ordering of document copies for delivery within two hours.

DSC charges for items supplied either as photocopies or loans. Lending is to registered users only, who are very often libraries. Individuals can register for a limited service, offering photocopies of copyright-cleared material, ordered by phone, fax or e-mail. The order has to include precise details of the item wanted, so it is necessary first to carry out a search to identify the source of information.

DSC is located at Boston Spa, in West Yorkshire, with easy access from Leeds, York and Harrogate. Its reading room is freely open to the public. Those who can visit can request any item from the collections for use in the reading room, although those wishing to consult more than five items are asked to give advance notice of five working days, after which their material will be waiting for them.

7.5.3   Public libraries

Public library provision varies greatly throughout the UK. Specialised scientific or technical information is available in the metropolitan areas and even at small service points there is generally advice on how the information can best be accessed, whether by borrowing material from larger collections, by telephoning specialist staff in larger centres or by contacting some other appropriate source identified from a directory.

The county library systems also have comprehensive technical collections; many have specialist staff responsible for helping with technical enquiries. These may be at a public service point or at the administrative headquarters, but in either case deal with any enquiries which cannot be answered by the local service point. It is increasingly common to have the technical and business material housed together, which is convenient for finding the different kinds of information needed during the innovation process; this can also be helpful when premises, and sometimes staff, are shared with the local Business Link or similar information centres. The growing network of Business Link offices is described in detail in Chapter 8.

If in doubt where to start in the public library system, make contact with the nearest library first. Where a specific item of printed information is wanted, and not available locally, public libraries can usually obtain it through a regional interlending system, or from the Document Supply Centre, the lending arm of the British Library. Libraries with small book collections are now often able to provide access to a wide range of materials via CD-ROM databases and online computer links.

People who work at a distance from their home address, or who commute to work, can normally become full borrowing members of the public library nearest their work. Addresses for all the public library systems in Great Britain are given in *Libraries in the UK and the Republic of Ireland*; some are mentioned on pages 153-4 in the country sections.

The Patents Information Network (PIN)

The Patents Information Network is a group of 14 libraries which helps local firms, inventors and the academic sector make use of the technical and commercial information contained in patent documents.

Twelve of the Network members are public libraries and they are joined by the University of Coventry Library and the British Library Science Reference and Information Service. All the libraries offer easy access to a collection of patent documents and related literature and their staff are trained to help enquirers find and make use of patents information.

The British Library Science Reference and Information Service in London holds the national collection of patents and plays a dual role in the Network. It acts both as the front-line Network library for the South-East and also as a centre of excellence to which other Network members can refer patents information enquiries, requests for database searches and photocopy requests.

The size of the patent collections varies according to local resources and needs, but all the libraries have staff who can show enquirers how to carry out a basic search of published patents and answer straightforward enquires about patents and the patents system. The range of services on offer will vary, but will include at least some of the following:

- access to Patent Office interactive training systems
- photocopying and document supply
- access to CD-ROM and online databases
- patent clinics
- meetings and workshops on patents and related topics.

The basic services, such as advice on using the patent collection and use of the Patent Office's unique interactive training systems, are free of charge, as are the Patent Clinics held in some libraries.

Photocopies will be charged for and the policy on other services (workshops for example) will vary.

Contact details for the Patent Clinics and the Patents Information Network libraries are given in Appendices 3 and 2 respectively, and more information is available on the British Library's Portico pages on the Internet's World Wide Web, at http://portico/bl/uk/.

England

Ten of the regional PIN libraries mentioned above are in England. In addition to their patents material they house the major technical collections in their area. At Coventry the collection is at the University of Coventry's Lanchester Library; the other PIN collections are in public libraries, in Birmingham, Bristol, Leeds, Liverpool, Manchester, Newcastle-upon-Tyne, Plymouth, Portsmouth and Sheffield.

These are all large municipal libraries with staff trained to deal with technical enquiries, and are worth checking by anyone in their area needing information. However, all public library service points are starting places for pursuing information needs.

Readers in the north of England should remember that the British Library's Document Supply Centre has a public reading room with access to a vast range of scientific and technical material for those who already have a reference they wish to follow up; details are given in the DSC section earlier.

Scotland

In Scotland, the main sources of scientific and technical information open to the public are in a small number of major centres. The Scottish Science Library (SSL), opened in 1989 in a purpose-built building in Edinburgh, aims to meet the needs of the scientific, industrial and business communities throughout Scotland. As part of the National Library of Scotland it benefits, like SRIS, from the legal deposit legislation and receives all new scientific books and journals published in the UK. It also holds the major abstracting and indexing services, CD-ROM databases and many of the publications mentioned specifically in the description of material held at SRIS. The Scottish Science Library collections are available for reference only, but admission is free to anyone on production of identification. Like many libraries SSL produces free guides giving details of the information available in particular subject areas of current interest or local strength; examples from SSL are leaflets on its extensive holdings of national and industry standards and a wide range of information sources on optoelectronics and artificial intelligence. For those who cannot visit, access to specialist help is available by letter, telephone or fax; further details are given in Appendix 2.

Scientific and technical information, including a lending service and help from specialist staff, is available at the central public library service points in Edinburgh, Glasgow, Aberdeen and Dundee. Glasgow and Aberdeen are Patents Information Network libraries and major centres for technical information in their areas, while some access to patents is also available at the Scottish Science Library and at Dundee. As in England, all public library service points can tap into a national network of information sources.

Wales

In Wales, the National Library of Wales at Aberystwyth does not have a separate technical library, but does benefit from legal deposit rights for all UK publications. The largest lending collection of scientific and technical material is at Cardiff Central Library. Its Scientific and Technical Section is open freely to all visitors for reference; those from outside the area may borrow for a small charge.

Northern Ireland

In Northern Ireland there is no legal deposit library, but Belfast has a large public library, which has a separate science library; it is a member of the Patents Information Network. In addition there is a regional network of libraries which, although small, actively promotes the use of technical information, channelling enquiries they are unable to answer to appropriate sources. Also, Queen's University, Belfast, is one of the universities which allows use of its collections for a fee to members of the public – this includes borrowing rights.

7.5.4 Co-operative information services

Public libraries are increasingly joining with other information providers in their area to co-ordinate a larger pool of resources and expertise. One example of several similar ventures is Hatrics, an information network of commercial and industrial companies, public libraries, academic institutions, government organisations and research establishments in Dorset, Hampshire and Somerset and surrounding areas. The services of information co-operatives like this are accessed through the member libraries and firms, so it is worth checking what is available locally. Small businesses are welcome to join, and may even be targeted specially. It is not always necessary to contribute to the pool of resources.

An increasing variety of information units are now offering a mix of free and priced services, and as the priced services are often open to anyone it is becoming easier for individuals to buy a piece of information in the same way as they would buy anything else they need.

Throughout the country there is a rapidly growing number of business information points supported by government as a source of help and information for local enterprise. Although focusing primarily on business information, some are developing initiatives in co-operation with local firms which extend to sharing technical know-how. In some areas both information resources and industrial workplace experience are shared, with the emphasis on helping people in the early stages of developing new technology, who will in turn help others. Expert advice for inventors is sometimes provided on the spot, and information on local resources is always available. In England the network of information centres is known as Business Links, in Scotland as Business Shops, in Wales as Business Connect, and in Northern Ireland as LEDU. More details of each are given in Section 8.7.7.

7.5.5 Non-public libraries

Access for members of the public to non-public libraries and information units varies greatly. Free help from academic or organisational libraries is likely to be limited to initial advice and brief queries, but it is increasingly common for organisations of all kinds to offer a mix of free and priced services, allowing free access to the users for whom the service was set up, but charging a fee to others. A variation on this is the free provision of basic services, with charges for specialised ones such as access to online databases or document copying. Large specialist libraries are increasingly open to non-members for a fee. If it is important to find a piece of information it is worth approaching any organisation in the subject area. Public libraries should be able to provide details of other libraries and organisations.

There is a detailed subject index in the *Aslib directory of information sources in the UK*, whose 8th edition has entries for over 8,000 organisations. Each entry has a brief description of the organisation, its subject coverage and special collections, publications, and any special information services. There is a note of who deals with enquiries, and in some cases an indication of how much help is available to non-members and whether there is a charge.

The British Library publishes two relevant guides: the *Guide to libraries and information units in government departments and other organisations* and the *Guide to libraries in key UK companies*. These guides also have subject indexes. Both are shorter listings than the Aslib directory, but concentrate on those which offer some degree of access to non-members. Some of these libraries have no restrictions on access, while others require a prior telephone arrangement or letter of introduction from a librarian explaining that other sources have been tried.

These directories will give ideas for pursuing a hard-to-find piece of information. Brief sections follow on the different kinds of non-public organisation as sources of information.

7.5.6   Universities

Because the attitude to access for the general public is both variable and changing, the only advice possible is to check with the particular institution. The situation varies from no access at all, which is fairly common because of shortage of resources for the institution's own readers, to an organised welcome for outside users, though almost always on a fee-paying basis. This can vary from a charge for reference or borrowing use of the university library, as at Belfast, to a priced special service aimed at helping local enterprise. It is always worth checking what is available locally and sometimes the initial exploratory discussion will be useful in itself, avoiding wasted effort.

At the time of writing the two university libraries which share legal deposit privileges with the national libraries, Oxford and Cambridge, do not offer any access to members of the general public.

Brief directory information for universities is given in the *Libraries in the UK and the Republic of Ireland* and more detail in the *Commonwealth universities yearbook* and *World of learning*. In any case a phone call to the main university library will establish the degree of access for non-members; those which allow access can usually send a brochure with details.

7.5.7 Government information units

Government libraries are seldom part of any formal information co-operative but some offer quite a lot of help to members of the public, as indicated in the British Library's *Guide to libraries and information units in government departments and other organisations*. The research institutions are sometimes prepared to help enquirers on a one-off basis, if staff are available. An example is the National Engineering Laboratory at Kilbride; while it does not formally offer provision of information to members of the public, its library and research personnel will on occasion try to answer queries when other available sources have been tried without success.

7.5.8 Independent and commercial organisations

As mentioned above, the British Library's *Guide to libraries in key UK companies* and the *Aslib directory* show the range of organisations which offer some help to individuals approaching them on a personal basis. Using the subject indexes of these directories and browsing through the entries can give leads to a huge amount of information which may not be readily found in more general libraries. To give one example, the library of the Institution of Electrical Engineers covers engineering, electronics, manufacturing engineering, computing, telecommunications, information technology and control engineering. The Library book catalogue can be found on the Internet at http://www.iee.org.uk, and loan services, for those who are not members, are available via interlibrary loan.

Professional bodies, such as the Institution of Electrical Engineers and the Royal Society of Chemistry, also often offer to search online information in their subject area as a priced service, as well as having a wide range of their own specialised publications.

The Institute of Materials provides a Materials Information Service (MIS) which offers a wide variety of information on engineering materials and processes. Resources used to answer enquiries include publications, contacts with materials expertise, and databases of trade associations and suppliers. The initial discussion of the enquirer's needs, and much information and advice, is available free.

It is worth noting that many of the professional and trade bodies listed in the above directories are the best source of information on the current regulations in their field and can often provide details of standard procedures and tests.

Approaching the appropriate specialised organisation is also a good way of finding expert help. Many organisations can put enquirers in touch with members who are prepared to help with the occasional query. There are also more formal services such as the British Cement Association's *Concrete helpline*, offering priced consultancy help, or *Concrete contacts* for free information on products and tradesmen. There are many directories which list trade and research associations; two of the most useful for the purpose of contacting experts are the *Directory of British associations*, and *Centres & bureaux: a directory of effort, information and expertise*. To locate further listings there is a directory of directories, *Current British directories*, which has over 4,000 entries. Each of these guides has a subject index to help identify the organisations in the field of interest.

For those who do not have the time or are unable to search for technical information themselves, there are independent information brokers who, for a fee, undertake research and locate information required by the inventor. They may use the publicly available resources described above, or have access to their own. Some of the independent brokers of services geared specifically to inventors' needs, such as product development, may be able to offer comprehensive help which can include finding technical information if necessary: some of these are discussed in other chapters. The possibility of expert help from the staff of libraries and other institutions listed in the directories above has been mentioned; journals for inventors, the *Yellow pages* directories and some daily newspapers all carry advertisements for individuals or small firms brokering information services. Their expertise should lead to efficient retrieval of the information if it is available, but it is advisable to examine their credentials and speak to satisfied users of their services before commissioning what can be a very costly search.

7.6   Online information

The following sections describe how to obtain information using a personal computer linked to remote databases:

- online searching of commercial databases
- scientific and technical information from commercial databases
- patents information from commercial databases
- the Internet.

The commercially operated services offering high quality information and sophisticated search techniques are mentioned first and then getting access to the many and varied sources available "for free" over the Internet is described.

7.6.1   Online searching of commercial databases

Definition
Online searching is a procedure by which you can search files (or databases) of information held on a commercially operated remote

computer (or host) from a terminal or microcomputer on your desk. The information in the files can be factual or numerical data, references to published documents (patent applications, journal articles, etc.) or the full text of documents. You pay for the information retrieved and for the telecommunications costs involved.

Why online?

An online search is quick. It is possible to achieve in an hour what would take several days of manual searching. Many inventors embark on a manual or CD-ROM search and then turn to online services when they realise how long it is going to take them and the effort they will have to make. It is also a very flexible option. On a specialist patent database for example, it is possible to look for patents:

- on a specific subject defined by keywords and/or classification codes
- linked to a known inventor or company
- filed with a particular authority
- filed over a given period of time.

Any or all of these approaches can be combined to create the kind of focused search that it is simply impossible to achieve using printed or microfiche indexes.

However, online searching has to be paid for. Unlike manual searching which can be spread across occasional days of research in a good patent or technical library, it is done in a concentrated burst of effort and there will be a bill at the end of the exercise. A supplementary manual search may also be needed if it is necessary to extend the search back beyond the early 1970s, which is the starting date for most online databases.

Getting an online search done and paying for it

The cost of an online search will vary according to which databases are selected and who does the searching. It is always difficult to estimate the cost in advance.

The best option will probably be to commission an intermediary (e.g. an information officer in a library or a patent agent/searcher) to do the searching and for the inventor to be closely involved in discussing the database and strategy options. If the inventor can sit in on the search, so much the better. If the search is to focus on patent databases, then the intermediary should have a good grasp of patent terminology and the patent system and should also be a frequent user of these specialist databases. Even seasoned online searchers can find it difficult to do a satisfactory patent database search and make sense of the data they retrieve. If the search is to cover general scientific and technical databases, it is again a good idea to select an intermediary who can give informed advice on the choice of databases and has some previous experience of searching them.

If an intermediary is commissioned to search, it is sensible to discuss charging policy, establish if VAT is or is not included in the prices quoted and establish what form the results will take before the work begins. It is *always* advisable to set a cost limit on the work.

Inventors tempted to carry out an online search on their own behalf should be fully aware of the cost and complexity of the exercise. DIY online searching (and especially DIY patent online searching) is an easy way of spending a lot of money very quickly, as it is a complex and expensive exercise. Any large public reference library or college/university library is certain to have staff who do online searching and anyone wishing to pursue the idea of DIY searching will be well advised to discuss it with them first. In practice, this DIY option will rarely be available to the independent inventor. Even if the necessary hardware and telecommunications connection are in place, the need to open an account with a database host system and acquire even the basic search documentation will be a substantial disincentive.

7.6.2 Scientific and technical information from commercial databases

The range of databases with information which might be of use in researching a technical topic is immense. It is estimated that around 1,000 databases might be used on some occasion, and around 400 more regularly. This means that it is important that any search service being used should offer a comprehensive range of databases and searchers familiar with the scope of the many files available, as the choice of the most appropriate database is crucial for a reliable and cost-effective online search. An intermediary with expertise in the subject area being searched is particularly useful when inventors have to investigate a topic outside their own field.

The following is a small sample of some of the databases likely to be useful for engineering enquiries:

- COMPENDEX PLUS for general engineering
- FSTA for food science and technology
- IBSEDEX for mechanical and electrical services in buildings
- INSPEC for electrical and electronic engineering, computing, and physics
- METADEX for the science and technology of metallurgy.

The most important database for chemistry is CA Search, the online version of *Chemical abstracts*, produced by the American Chemical Society. It has over 12 million records, covering all aspects of chemistry and related sciences, and includes patents in the sources of information it searches. Just as the non-patent literature should be remembered when investigating the novelty of an idea, the patent literature should not be overlooked as a source of detailed, illustrated technical information.

7.6.3   Patent information from commercial databases

There are two groups of databases to be aware of as sources of patent information:

- patent databases, which provide detailed information on published patents/patent applications
- science and technology databases, which cover a range of information sources such as journal articles and conference papers (and sometimes patents as well).

The science and technology databases should be considered as a possible source of information if an extensive prior art search is needed, since an invention might have been disclosed in a journal article or conference paper without ever having been the subject of a patent application.

The most widely used of the specialist patent databases are World Patents Index and INPADOC (see below). Many others are available, most of which cover the patents issued by a single Patent Office, but there is no online database which only covers patents applied for or granted in the United Kingdom. UK patents and European patents designating the UK are, however, covered on both World Patents Index and INPADOC.

The World Patents Index provides information on granted patents and published applications and now covers the activity of over 40 patent offices worldwide. Coverage of all patentable technologies started in 1974 and some technologies (e.g. chemistry) have been covered from the mid 1960s. All the major patent offices are covered, although Japanese coverage was limited to chemical patents and those in the electronics/engineering fields until 1994. During 1995 coverage was expanded to cover Japanese patents in all areas of patentable technology. The majority of patent records on the database include both an expanded English-language title and a comprehensive summary of the invention described in the patent specification. It is possible to search the database by names (inventors or companies), dates, patent numbers and country of origin. The records can also be searched by subject. This means that it is possible to search for patents on a particular topic using English-language keywords, codes from the International Patent Classification which cover the topic and codes from the database's own classification scheme for patentable technology. The database is produced by Derwent Information Ltd.

INPADOC provides information on granted patents and published applications from over 50 patent offices. Basic coverage of UK patents began in 1969 and the amount of detail available was gradually increased, although inventor names were not added until the early 1980s. European patents have been covered since the system was introduced in 1978. The database can be searched by names (inventors or companies), dates, patent numbers and country of origin. It is also possible to do a search for patents on a specific topic by using International Patent Classification codes, but keyword searching is not an option on INPADOC. The records consist of a patent title (in the original language) and associated names, dates and numbers. The database is now produced by the European Patent Office.

7.6.4   The Internet

What the Internet offers the inventor
A computer, a modem and a telephone line give access to the Internet, a world-wide system of interconnected computer networks. More detail on equipment and cost is given on page 168 in the section "Setting up a personal Internet connection". From the huge range of resources available, those most useful to the inventor are personal electronic mail, discussion lists and newsgroups, and the World Wide Web.

Electronic mail
Electronic mail is used as an alternative to post or telephone, combining the advantages of both, giving low-cost, rapid communication with a written copy. Many professionals and organisations can now be contacted in this way.

Discussion via the Internet
Mailing lists and newsgroups give access to people who are willing to help with specialist information and advice. Information may be obtained by a question addressed to a known person, or by sending an open question to all those who participate in an appropriate subject-based discussion group. Before sending a question to the whole group, it is advisable to check whether it has already been answered. Most groups maintain lists of "Frequently asked questions" (FAQs), and these are the best starting points.

There is a vast number of these discussion groups. Some of the many relevant to the inventor include:

- alt.inventors (discussion of inventions)
- alt.int-property (all types of intellectual property)
- sci.engr (engineering in general)
- sci.engr.manufacturing (manufacturing technology)

- sci.engr.mech (mechanical engineering)
- sci.materials (all aspects of materials engineering).

There is a continuous flow of new contributions to these groups, each of those listed receiving about 100 items per week at the time of writing. These contributions include questions and answers from all over the world, contributors ranging from amateur enthusiasts to professional engineers from universities, industry and government. They share an interest in finding solutions to problems and exchanging experience. Discussions cover basic principles as well as sources of supply and evaluation of specific products and materials. Examples of typical interchanges are shown in Figure 10.

The World Wide Web (WWW)

The WWW is a network of information resources accessible via the Internet. These resources include documents, directories, and images, each of which can contain links to other sites on the Web. By following these links the user can immediately and automatically be connected to the other resources, wherever in the world they happen to be held. The information found can be details of organisations, usually compiled by themselves, or information about a subject. It is increasingly common for sites to give access to full documents, some already published elsewhere, some available only from the Internet site. The advantages of publishing information in this form include speed and cheapness for both providers and users of the information.

Finding information on the Web

Answers to specific subject questions on the Web can be sought in two different ways. There are lists of resources grouped by subject. Working through these can lead to sites in which relevant information may be found by browsing. The World Wide Web Virtual Library <http://www.w3.org/vl/> is a good starting point to find these lists; another is <http://www.yahoo.com/>. Engineering information specialists at Heriot-Watt University Library manage a database <http://www.eevl.ac.uk> which has links to over 1300 Web resources, with engineering topics search-

Figure 10. Discussion on the Internet: excerpts from a typical exchange from the mechanical engineering group.

Subject: Re: Cylinder under internal/external pressure

[The original query was: "My understanding is that a cylinder can withstand a greater pressure exerted internally than externally. My question is 'are the stresses the same in both cases, but the allowable stress greater for a cylinder under external pressure?' Is the formula for calculating stress in a cylinder subject to vacuum different from one subjected to internal pressure?"]

Replies included:

```
14 Apr 1995
From: prime@lanl.gov (Mike Prime) at Los Alamos National Lab

  The stresses in a cylinder under vacuum can be calculated
with the same formulas as for a cylinder under pressure. If
you follow the derivations you will see how they work for -P
(works for both thick and thin wall formulas). In general you
don't do this because the failure mode is generally buckling.
But occasionally it does come in handy. For example you can
calculate the residual stresses from yielding a thick walled
pressure vessel by elastically unloading (negative P).
Mike


  15 Apr 1995
From: mstcheung@alpha.ntu.ac.sg at Nanyang Technological
University — Singapore

  In the design of hollow cylinders subjected to internal
pressure, the pressure is limited by plastic yielding of the
material if it is ductile, and by ultimate fracture or
rupture. Just equate the hoop stress of pr/t to the allowable
stress based on yield and tensile strengths.
As for hollow cylinders under external pressure, it depends
on how thin are the cylinders. Very thin cylinders tend to
buckle elastically. How big is your cylinder, and how high is
your pressure?
The ASME has a pressure vessel code, called Section VIII,
Division  1,  which  provides  procedures  for  designing
pressurized cylindrical vessels to withstand both internal
and external pressure.
On the other cylinder, you need to look into designing the
ends, and the closure.
Dr John Cheung .


  18 Apr 1995
From: s10stewart@aol.com (S10Stewart) at America Online, Inc.

  There are many design books that address this subject but
what is used mostly in industry is ASME section VIII Division
I and II "Pressure Vessel Code"...
```

able by title, keyword or subject. The British Library site, <http://portico.bl.uk>, is another online starting point, and the Patents Information Web pages on Portico include information on other patents sites.

On the other hand there are computer-compiled indexes of words found in headings and documents, which can be useful, but these have the limitations of any free text searching. The different indexes have some overlap, but their scope is difficult to determine, so although it is worth using several of them the results cannot be relied upon as exhaustive.

It is important to realise that the Internet does not contain anything like the detailed and comprehensive coverage of published material which is found in commercially-produced indexing and abstracting services and their online database versions, and therefore cannot be a substitute for them. It is, however, complementary, as it gives access to material which has not yet appeared (and may never appear) in published sources. Much of the information found on the Internet is informal, and as with any other source of information its authoritive nature should be assessed by considering who has written it, the data on which opinions are based, whether it is agreed and confirmed by other people, and whether the author is speaking personally or on behalf of an organisation.

As well as being a source of organisational and subject information, the Internet offers the chance to take part in activities with others who share an interest, for example The Engineers' Club. This calls itself a technically-focused online service, and although aimed at the engineering profession could be a way for those with a spare-time interest to keep in touch with relevant activities.

Although for the present and in the near future most of the information needs of the inventor will best be met by conventional methods, the amount and range of data available on the Internet is increasing all the time. Already it is a valuable further resource of unpublished data and contacts with researchers working in a specialist area to share problem solving. Anyone who has access to the Internet, and has experience in searching for subject information, should check what is available.

Anyone not already using the Internet would probably be advised to look for their information via other routes for the present, but those who want to investigate further will find some basic information in the next section. A taste of what Internet searching involves before setting up a personal connection can usually be had somewhere in your area, in the increasing number of public libraries and other information services which offer access, or in informal "cyber cafés", which offer search sessions in return for a fee. To get an idea of the range of material available before going online consult *The Internet for scientists and engineers*, now in its second edition, and use the suggestions at the beginning of this section.

Setting up a personal Internet connection
Read a general introduction to the Internet
There are many guides for beginners, and reading one of these would give a more detailed idea than is possible here of what is involved. Appendix 8 gives a few titles, but the information is so quickly superseded that it is best to check the local library and bookshop for the most up-to-date titles. Once a decision to go ahead is made, much useful information is available free on the Internet itself. It is a good idea to read some recent issues of periodicals devoted to Internet topics, which include advertisements for hardware, software and support services.

Check the hardware and software requirements
Most modern computers can send and receive Internet mail and news, as well as text files and documents. Software is available for IBM compatible, Apple and Unix machines, and for many others. The other piece of equipment needed is a modem, to connect the computer to the telephone line. To receive online images without undue delay a minimum speed of 14,400 bytes per second is necessary and, if a new modem is being bought, it is worth the small difference in price to get one which runs at 33,600 bytes per second. To use the World Wide Web, with pages which may contain images, the minimum recommended processor is a 486DX, 33 MHz, with 8 Mb of memory and a Super Video Graphics Array (SVGA) screen.

The minimum programs needed to get started are those which send and receive mail and transfer files from other computers; suitable software may be supplied by the service provider. Once a connection is made, other software is obtainable from the Internet itself, some of it free or "shareware" (i.e. available for a small fee after trial use). For example, a graphical "browser" interface is needed to see World Wide Web pages laid out properly with images and tables. Commercial software packages and manuals are available which provide an integrated collection of many of the most useful programs; these may help to reduce the problems of setting up a connection, which can be quite tricky when configuring several different pieces of hardware and software.

Find a service provider

For those who do not have access to a government or university computer network, commercial organisations can provide a connection to the Internet for a fee. Prices vary, and can include an initial setting-up charge, monthly or annual fees, and charges for connect time or amount of data sent. A typical charge in the UK at the time of writing for a dial-up connection of a single computer is about £120 per year, with telephone call charges being paid to the telephone company in addition. Call charges are to the nearest connection point, and as these points are available throughout the country local rates usually apply; the fact that you may be connecting to computers on the other side of the world does not increase the cost. It is often possible to send and receive mail and news quickly during a brief connection and then to disconnect so that there is no telephone charge to pay for the time taken to read and reply to messages. Telephone costs mount up when using on-line services if overload of computers or communications links causes delays in transmission; it is therefore best to use this type of service in the mornings, before users in the USA become active.

A selective list of some of the current service providers is given in Appendix 6.

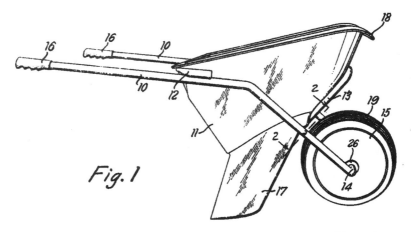

Fig. 1

Figure 11. British patent number 1510011. James Dyson's
Ballbarrow wheelbarrow patent, applied for in 1975.

8
Sources of Business Information

Nigel Spencer

Business information sources can be used to inform many of the decisions the inventor has to make in developing an invention. Examples include determining the market for a product; locating suppliers and manufacturers; assessing companies as licensees and sources of funding; or searching for customers. This chapter describes the key sources and how to access them for yourself.

The chapter is in two distinct parts:

- Sections 8.1–8.6 list published sources of information on all the main business topics of interest to inventors:
  - company financial information
  - company ownership and personnel
  - company activities
  - market information
  - information on new and existing products
  - how to find out what sources of business information exist.
- Section 8.7 gives guidance on locating and gaining access to the data listed in 8.1–8.6, and describes the main sources offering professional help:
  - The British Library Business Information Service
  - The British Library Lloyds Bank Business Line
  - The Business Information Research Service
  - The Export Market Information Centre (EMIC)
  - The Chartered Institute of Marketing (CIM)
  - public library services

- other business information services
- business information on the Internet.

8.1 Company financial information

This section is prefaced with a word of warning. First, all financial information contained in company accounts is historical: a company's financial health may have improved or deteriorated substantially since the close of its most recent accounting period. Secondly, a company's duly audited accounts may not of themselves convey helpful or accurate information. This is both because accounts may be prepared in accordance with different accounting conventions and because they are prepared in relation to financial data which is provided by the company to the accountant, the reliability of which cannot always be ascertained by the audit process.

All publicly available financial information on companies is based on the accounts published by the companies themselves; in the case of UK companies these are deposited at the Companies Registration Office otherwise referred to as Companies House. The financial data quoted in the publications and services listed below will, therefore, be derived from the same data source. What will vary is the way that this data is presented, the figures that are included and the extent to which it is manipulated and analysed.

8.1.1 Companies House, company annual reports and interim reports

Company annual reports and interim reports are the most obvious sources of financial information on companies. Company accounts will be presented in full with a set of accompanying notes. Reports will often be supplied by companies on request. In the case of UK companies, accounts for all limited companies can be obtained by visiting Companies House in London or contacting the Companies Registration Office in Cardiff.

Companies House directory of British companies
This is a microfiche directory produced by Companies House which provides some essential information on all limited companies. It states when the most recent accounts and returns were deposited at Companies House, when the next returns are due, the company registration number and registered office address. The directory is published quarterly with weekly updates.

ICC business ratio reports
This is a series of over 200 published reports, each of which focuses on the financial performance of companies operating in a particular sector of UK industry. Each report contains a financial profile of every company featured as well as a range of ratio tables which enable users to compare the performance of companies by a variety of financial criteria. This enables researchers to compare the financial performance of companies of different sizes. Each report contains data on between 80 and 150 companies.

8.1.2 Trade directories

Trade directories seldom provide detailed financial data on companies. However, some can be useful in locating key figures, such as sales or profits, which will indicate the size of companies. When looking for data on smaller UK companies, trade directories may provide the only source of information without having to contact Companies House. The publications listed below are some of the most useful.

Kompass UK
The Kompass publications are a series of directories covering over 25 countries in Europe, Asia, the Middle East and the Pacific Rim. They provide a good coverage of companies in these countries with an emphasis on manufacturing businesses. The amount of financial data given varies from country to country. There is a separate financial data volume for the UK which provides basic financial data for the previous three years; this volume also provides names of parent companies and major subsidiaries. Data on most other countries is less detailed, usually limited to a figure

for share capital or sales. The names of key directors or contacts are provided for all countries. *Kompass UK* is published by Reed Information Services.

Key British enterprises (KBE)
This is a six volume directory which provides profiles of the top 50,000 UK companies. The data provided includes the latest figures for sales, profit/loss and authorised capital. In addition the names of some directors are listed. The directory also provides indexes by area of activity, location and export markets as well as a trade name index. Published annually, it is produced by Dun and Bradstreet.

Hambro company guide
This provides profiles of companies with full Stock Exchange listings, as well as those quoted in USM and OTC. Each profile contains five years' key financial data, and some provide details of share performance. In the case of very large companies, the profiles provide names of directors as well as extracts from the text sections of the companies' annual reports. It is published by Hemmington Scott Publishing Limited and is produced quarterly.

Macmillan's unquoted companies
This is a two volume directory which provides financial profiles of the top 20,000 companies in the UK. It often provides data on companies not covered by other directories. The profiles also give names of key directors and parent companies. It is published annually by ICC Business Publications.

8.1.3   CD-ROM services

CD-ROM products provide an excellent means of obtaining financial information on companies. Researchers can use them not only to find information on known companies but also to search on a range of different criteria, financial or otherwise, to identify potentially interesting companies. The CD-ROM format is becoming increasingly popular with publishers and with users. Because large amounts of data can be stored on CD-ROMs, researchers find that they can consult just one disc and not several printed volumes as they may have needed to do in the past. Some of the most useful products currently available are listed below.

FAME
Provides detailed financial data on 95,000 UK companies. It is produced by the CD-ROM Publishing Company and is based on data compiled by Jordans. It is updated monthly.

International public companies (EXTEL)
Provides descriptive information and five years' worth of very detailed financial information on the top 10,000 public companies worldwide. Updated monthly, it is produced by One Source and contains data compiled by Extel Financial. Extel data can also be accessed via the Extel Financial workstation.

Disclosure
Disclosure produce two discs. *Disclosure United States* provides detailed financial data plus text from annual reports on 12,000 US public companies. The second, *Disclosure worldscope*, provides financial data on 6,000 top companies in 25 countries. Both disc services are updated monthly.

AMADEUS
Provides financial data in profiles of 130,000 companies throughout Europe. It is produced by the CD-ROM Publishing Company and is updated bi-monthly.

8.2 Company ownership and personnel

Information on the names of directors within companies, or the names of subsidiaries or parent companies, can be found in many of the sources of company information described above: annual reports, company information CD-ROMs or directories. In addition three other specialist sources can be consulted.

Who owns whom

This directory specialises in company relationships and is produced annually by Dun and Bradstreet. It is published in a number of volumes, each relating to a specific geographical area: UK, Continental Europe, North America, Australasia and the Far East. Each volume allows searching by the parent or the subsidiaries of a particular company.

Directory of directors

As the name suggests, this directory provides a listing of company directors. It covers the top 15,000 companies in the UK and is published by Reed Information Services. The directory consists of two volumes. Volume 1 is an alphabetical listing of directors which enables the researcher to check which companies that individual is linked to. Volume 2 is organised by company name and provides a list of directors for each company. It is published annually.

Arthur Andersen corporate register

A publication which is similar to the *Directory of directors* in layout. It covers only 2,200 UK quoted companies but provides a greater level of information on those companies. Entries consist of a profile of the company with names of executive and non-executive directors, as well as information on total directors' pay. The *Register* also includes an alphabetical listing of individual directors' names with brief biographical details and a list of the companies with which they have affiliations. It is published twice yearly by Hemmington Scott Publishing Ltd.

8.3 Company activities

The information sources providing information on company activities fall into two distinct categories. Firstly, there are information sources produced by the companies themselves such as annual reports and house journals. Secondly, there are information sources such as trade journals and newspapers which are produced by organisations independent of the companies about which they write. It is important to be aware of this distinction when carrying out research. Few companies will want to stress their less successful activities. It is generally the case that the larger and more significant the company, the more will be written about it. Therefore, when researching smaller companies it is less likely that they will be mentioned in sources such as the national press. It is also unlikely that a smaller company will produce a glossy and informative annual report or a company house journal. Researchers may find that the only possible source is the specialist trade press since it focuses on the industry in which the company operates.

8.3.1   Annual reports

Annual reports for larger companies will contain passages of text such as the chairman's statement or organisational review which will review the activities in which the company has been involved over the past year and discuss future developments.

8.3.2   House journals

Company house journals are primarily aimed internally, at staff within the company. Many will contain articles about the ventures in which the company is involved. The quality of this kind of publication is highly variable and the level of content of potential interest to the researcher can be limited. However, where such information does exist it can be particularly valuable as it is unlikely to be found anywhere else. Limited circulation house journals may be the subject of obligations of confidentiality and it cannot be assumed that the information contained in them is freely available.

8.3.3   Trade journals

Trade journals covering a particular industry are an excellent source of information on companies operating within that industry. This information can take the form of stories about new products and ventures, features on personnel within the company or news about the difficulties which a company is experiencing. This information often appears in the form of short, one paragraph news items. In the case of companies which are key players in their industry, there is an increased chance of finding longer, more analytical and detailed articles. It is seldom the case that trade journals do not provide any company information.

8.3.4   Industry newsletters

These publications tend to present information in a more condensed style than in trade journals. This is because they are designed to be scanned quickly by persons needing to keep abreast of developments within a particular industry. They will often be the first publications that report new product developments or important company news. This type of publication is particularly strong in fast moving industry sectors such as healthcare, pharmaceuticals and information technology. All industry newsletters will provide company news in some form.

8.3.5   Newspapers

National broadsheet newspapers often publish company news in their business sections. This information will usually cover only large domestic or multinational companies. Stories may also mention companies which are significant players within their own industries. *The Financial Times* and *The Wall Street Journal Europe* have a particular focus on business news and are perhaps the key titles; however, *The Independent, The Independent on Sunday, The Daily Telegraph, The Sunday Telegraph, The Times, The Sunday Times, The Guardian* and *The Observer* also carry a significant amount of company news.

8.3.6   McCarthy press cuttings

The McCarthy services provide press cuttings on individual companies, allowing researchers to scan quickly for news of key developments in particular companies. The sources of the cuttings are newspapers and selected trade and business journals. The

McCarthy services are divided as follows: UK quoted, UK unquoted, European and Australian. McCarthy also run a service covering individual industries or topics as varied as funeral directors, privatisation and herbicides. These services are updated on a daily basis.

8.3.7 CD-ROM services

When searching for information on company activities CD-ROM services are an excellent source combining speed with precision. The following services are particularly useful:

McCarthy on CD-ROM
CD-ROM version of the press cuttings service produced by McCarthy. The sources covered are over 50 newspapers and journals. Each disc provides six months' coverage. The service is updated monthly.

Newspapers
Full text of newspapers are increasingly becoming available on CD-ROM. Two that provide very good business coverage are the *The Financial Times* and *The Independent*. Both are updated quarterly. *The Guardian, The Daily Telegraph* and *The Sunday Telegraph* are also available in CD-ROM format. Some newspapers also have World Wide Web sites, including *The Financial Times, The Telegraph* and *The Times*.

8.4 Market information

An essential factor in the commercial exploitation of any invention is to establish the nature of the market which the product would enter. There are a number of published sources of information which can help you to get a feel for the size of a market, whether it is in the process of growth or decline and which companies currently have a major share of that market. The major categories of published information are listed below along with details of publishers and names of key sources. Information sources on consumer market information and industrial market information are treated separately. The reason for this is that market research on consumer goods is comparatively much easier to find, so the methodology involved in finding consumer market research differs from that needed to locate comparable information on industrial sectors.

8.4.1 Consumer market information

Consumer market information is easier to locate because there are a large number of companies producing off-the-shelf consumer market reports at prices considerably lower than those charged for similar industrial market reports. Libraries are therefore likely to hold more comprehensive collections of consumer reports.

Consumer market research reports
Market research reports are the principal source of market information on consumer goods. The data contained in these reports is a combination of information taken from official government statistical sources and unofficial statistics produced by bodies linked to particular industries. Some publishers will also undertake some original research of their own. A number of publishers lead the field in the production of UK consumer market research, but many cover industrial sectors as well and some do extend coverage beyond the UK.

The major publishers of off-the-shelf consumer market research reports are listed below. In addition the list includes some examples of publishers which specialise in particular sectors. (This specialisation is more common in industrial rather than consumer market research.) Reports by these specialist publishers are always worth consulting when researching a very specific sub-sector or niche market.

Key Note
Key Note produce a series of reports covering a wide range of sectors which provide a very good starting point for research. There are currently over 230 reports in the series covering general sectors such as household goods, leisure, printing/publishing,

retailing, tourism, drinks and tobacco, as well as more industrial areas such as engineering, chemicals and transport. These reports provide a general overview of the market in question, with data on market trends, key players and a useful list of other information sources such as trade journals and associations.

Market Assessment

Market Assessment publish a series of very detailed reports on UK consumer goods. The general categories covered are business services, drink, electrical, financial, food, home, leisure and personal care. There are currently over 70 reports in the series with over 30 titles being published a year. The reports contain sectoral overviews as well as individual reports providing data on segments of that market. Researchers looking for information on a particular sector where a Market Assessment report is available are likely to find it provides most of the data they require.

Mintel

Mintel is a well known publisher of consumer market research. This research is published in the form of special reports and four journal titles. The four journal titles are:

- *Market Intelligence*: six reports per issue, published monthly. Reports 20–30 pages.
- *Retail Intelligence*: two reports per issue, published bi-monthly. Reports 50–100 pages.
- *Personal Finance Intelligence*: four reports per issue, published quarterly. Reports 20–30 pages.
- *Leisure Intelligence*: six reports per issue, published quarterly. Reports 20–30 pages.

Market Intelligence covers the widest range of consumer products and services, with *Retail Intelligence* looking at the retail sales aspects of certain goods and services. The coverage of *Personal Finance Intelligence* and *Leisure Intelligence* focuses on products and services relating to these sectors.

Mintel also produce around 20 special reports each year. These provide more in-depth analysis of a wide range of consumer markets as well as focusing on the lifestyles of certain age groups.

Euromonitor

Although involved in a much wider range of publishing activities, Euromonitor produces a significant number of reports on consumer markets. It publishes over 140 reports on a number of broad areas: alcoholic and soft drinks, automotive, book publishing, catering and vending, domestic electrical appliances, food, healthcare, household cleaning, household products and furnishings, leisure, lifestyles, packaging, personal finance, retailing, tobacco, travel and tourism. The reports are international in coverage, either covering the world markets or focusing on a region such as Latin America or Asia. It is this international coverage that is the particular strength of these reports. Worldwide market reports are usually part of the Market Direction series.

- consumer market research is also published by Euromonitor in four monthly journals:
  - *Market Research Great Britain*. Each issue contains six sector reports.
  - *Market Research Europe*. Each issue contains two pan-European sector reports and four sector reports covering individual countries or regions.
  - *Market Research International*. Each issue contains worldwide sector reports and four sector reports on countries outside Europe.
  - *Retail Monitor International*. This provides retailer profiles, looks at current trends and the outlook for various retail sectors. The coverage is worldwide with reports focusing on individual countries.

In addition Euromonitor produce a series of European and worldwide reference books. These fall into the following categories: marketing directories, directories of information sources, regional handbooks and statistical publications. Some of the most useful of these for those looking for consumer market information are:

- *International marketing data and statistics*
- *European marketing data and statistics*.

These contain economic and socio-economic, trade, demographic, travel and health statistics. They also provide consumer market size data for individual countries, but only for major market sectors:

- *Consumer Europe*
- *Consumer Japan*
- *Consumer USA*
- *Consumer Eastern Europe.*

These reports contain a considerable amount of market data on major consumer market sectors within the regions covered, and in the individual countries within those regions. The information takes the form of numerous tables with virtually no commentary or analysis.

Datamonitor
Datamonitor are prolific publishers of over 130 market research reports covering the UK, continental Europe and to a much lesser extent the USA.

Business monitors
Quarterly and annual series of statistics from the UK Central Statistical Office, and published by The Stationery Office, covering the majority of UK business sectors. The individual reports are classified on the basis of Standard Industrial Classification (SIC) (1980) codes. The reports include sales, import and export figures. Many published market research reports contain figures taken from Business Monitors.

8.4.2 Industrial market information

Finding market research on industrial sectors is more problematic than finding similar information on consumer market sectors. This is largely because where published research does exist it is usually very expensive and this limits its availability. Researchers will often find that they have to consider more accessible alternative sources such as trade journals, newsletters, CD-ROM products and abstracting journals.

Industrial market research reports
Specialisation is common amongst publishers of industrial market research. With a few exceptions publishers focus on a particular industrial sector, or related sectors. The publishers listed below either cover a range of sectors, or are good examples of specialists.

Frost and Sullivan
US-based publisher of expensive reports on industrial sectors. The reports cover industries undergoing frequent changes brought about by technological developments. The geographical coverage is either US or European and the subject coverage of each report is often very specific. Examples of some of the broad industry sectors covered by Frost and Sullivan are telecommunications, chemicals, healthcare, information technology, advanced materials, plastics, electronics, engineering, waste management and defence. Prices per report range between $1000 and $4000.

Economist Intelligence Unit
This publisher produces a number of market research publications, most of which cover individual industrial areas but some cover a range of sectors. The areas of specialisation include the automotive industries, textiles, paper and packaging, rubber, financial services and travel. The geographical coverage of these publications varies from worldwide to regional surveys, to reports on individual countries. The research is published in the form of special reports and journals. The journals provide condensed and up to date information and are an excellent source for recent data.
   The multi-sector journal titles are:

* *Retail Business*: published monthly, each issue contains reports on four UK consumer goods markets
* *Retail Trade Review*: published quarterly, each issue contains statistics and forecasts for UK retail sectors
* *Marketing in Europe*: published monthly, contains studies of consumer markets in France, Germany, Italy, Belgium and the Netherlands with occasional coverage of other countries.

The most useful industry specific journal titles are:

- *European Motor Business*: published quarterly
- *International Motor Business*: published quarterly
- *Japanese Motor Business*: published quarterly
- *Travel and Tourism Analyst*: six published annually
- *International Tourism Reports*: published quarterly
- *Paper and Packaging Analyst*: published quarterly
- *Rubber Trends*: published quarterly.

Financial Times Business Information
The Financial Times produces a series of reports entitled Management reports on a range of topics. The broad sectors covered are accountancy, automotive, banking and finance, energy, insurance, marketing, packaging, pharmaceuticals, property and telecommunications. The geographical coverage varies, with many reports providing international coverage.

Marketing Strategies for Industry
MSI is a UK publisher which has, in recent years, shifted subject emphasis from consumer goods to more industrial and specialist sectors, while remaining a significant producer of consumer market research. It publishes over 150 reports on subjects as diverse as powered access platforms, Indian food, urban development corporations and waste management. The majority of reports cover the UK only, but increasingly MSI is publishing reports covering other Western European countries.

Elsevier Advanced Technology
As its name suggests, this publisher specialises in reports on industry sectors which are at the cutting edge of technological development. Electronics and electronic components are a particular specialisation but other sectors covered are plastics and telecommunications. The majority of these detailed reports are worldwide or European in coverage. The price of reports can be as high as £1,300.

Ovum

Ovum publishes market reports covering leading edge technologies in the areas of computing and telecommunications. These reports provide in-depth coverage of very specialised topics within these broad sectors. The reports are priced in the region of £1,200.

Trade journals

Researchers looking for market data on a particular industry will find it is always worth investigating any trade journals covering that industry. This is because these journals sometimes provide short summaries of the key data contained within expensive published reports as well as including unofficial statistics collected by trade associations. It certainly is not the case that all trade journals will provide market data but a significant proportion will. Some examples of trade journals which contain market data are:

- *International Cement Review*: published monthly by Tradeship Publications
- *Metal Bulletin Monthly*: published by Metal Bulletin PLC
- *Chemical Week*: published weekly by Chemical Week Associates
- *Telecommunications*: published monthly by Horizon House Publications.

Industry newsletters

These are a good source of official and unofficial statistics and market data. In fact, statistics often appear initially or uniquely in industry newsletters. They are also particularly readable, presenting information in a condensed form which allows quick scanning. The industries that are well covered by this kind of publication are information technology, pharmaceuticals, energy and the media. Some examples of useful newsletters are:

- *Scrip* (covers pharmaceuticals): published twice a week by PJB Publications
- *New Media Markets*: published fortnightly by *Financial Times Business Information*

- *Blackwell's Software Markets*: published fortnightly by Blackwell Professional Information Services
- *Infotecture Europe* (covers the information industry): published fortnightly by A Jour
- *Power in Europe*: published fortnightly by *Financial Times Business Information*.

Many trade journals or newsletters do not provide their own indexes but an alternative to scanning these volumes is to use an abstracting or indexing journal or a CD-ROM search. When searching for information on market sectors or industry activities CD-ROM services provide the opportunity to search very quickly and comprehensively. By simply inputting a search term you can retrieve all mentions of that company contained on the disc. Searches can then be further refined by the inclusion of additional search terms. The following service is a useful source of market data particularly on industrial sectors.

F&S Index Plus Text
This service provides references to articles in trade and business journals published worldwide. Eighty per cent of the records contain either the full text of the article or an abstract. There are two discs: a US and an international (non-US) disc. Industries can be precisely searched using the detailed Predicasts product codes (these are based on the US Standard Industrial Classification (SIC) Scheme with each code relating to a particular industrial sector). In addition Predicasts use a set of event codes which enable the researcher to identify articles by the type of information they contain, such as market information or information on new products. The service is updated monthly and is produced by Information Access Company.

8.5 Information on new and existing products

Information on other products is of value to an inventor as it will establish what existing products may compete with the idea that they are developing. This information could alert them at an early stage to the fact that their idea is not novel and thus save a considerable amount of time. This information could also help them to develop products that had unique features by identifying the gaps or deficiencies of existing products and product ranges.

Below are listed a number of valuable sources of information for researching product information and monitoring new product developments.

8.5.1   Trade literature

Trade or "product" literature is material produced by companies providing data on their products. It can be found in the form of catalogues, leaflets, datasheets or press releases. The form and the kind of data provided is usually dictated by the nature of the product concerned. Because it is aimed at potential purchasers the data provided is mainly product information written to persuade customers to make a purchase or request more information from the company. There are few accessible library collections of product literature, but the British Library Business Information Service and the Science Museum Library both have collections. It should be remembered that literature can be obtained by contacting companies direct.

8.5.2   Journals and newsletters

Specialist periodicals of this type are a very good way of keeping up to date with new product developments and virtually all titles provide product information at some level. This usually takes the form of new product sections which report on new product launches or significant new product developments. Trade journals also often contain enquiry services which enable readers to request detailed data on new products. In the case of industries, such as the pharmaceutical industry, where licensing and testing are crucial in the progress of a new product, specialist newsletters play an important role in keeping readers informed of key developments.

Less common, but particularly useful, are product range surveys which sometimes appear in trade journals and newsletters. These allow the price, performance and specifications of competing products to be compared.

Journals which focus specifically on product information are a somewhat disparate group. They tend either to look at new products or provide surveys of product groups which enable readers to compare existing products. The most common kind of publication covers a particular overall product group such as cars, videos or personal computers and is aimed at the informed consumer. Some with a broader subject coverage are listed below:

What to Buy for Business
A monthly consumer report which covers business equipment and services available in the UK. Each issue contains between one and three detailed surveys all of which contain recommendations and useful features such as glossaries of terminology. Examples of the kinds of products covered are matrix printers, temp agencies and air conditioners. Published by Reed Business Publishing.

Which?
A well known and often quoted monthly journal providing surveys of UK consumer products and services. Published by the Consumers' Association.

What's New in Industry
A monthly journal which provides brief details on new products under a range of general headings such as instrumentation and control or hand tools. Every product is allocated a number and, by entering this number on an enquiry card and posting the card, you can receive further information on the product. Published by Morgan Grampian PLC.

8.6   How to find out what sources of business information exist

Previous sections have looked at what categories of publication can be used to find certain types of information. A number of individual publications have been mentioned, but these are only a small part of what is actually available. To get a more comprehensive list of the publications available on a particular subject indexes and directories should be used. Some of the most comprehensive of these are listed below.

8.6.1 Directories of trade directories

Current British directories
Probably the best and most comprehensive list of UK directories. Published every two years by CBD Research Ltd.

Directories in print
A comprehensive directory with worldwide coverage. The 1994 edition has entries for 15,900 directories. Published annually by Gale Research Inc.

8.6.2 Directories of market research reports

Marketing surveys index
A cumulative monthly looseleaf directory which covers published market research worldwide. Published by Marketing Strategies for Industry.

Marketsearch
An international directory of published market research. The 1993 edition had entries for 18,000 reports. Published annually by Arlington Management Publications.

Findex
A worldwide directory of market research reports and surveys. The 1993/94 edition had entries for 12,000 publications. Published annually by Euromonitor.

8.6.3 Directories of journals and newsletters

Benns media
An international directory of newspapers, periodicals, television and other media. It comes in three volumes; UK, Europe and World. Published annually by Benn Business Information Services Ltd.

Ulrich's international periodicals directory
Covers 126,000 periodicals published worldwide. Published
annually by R.R. Bowker Co.

8.6.4 Directories of CD-ROM products

The CD-ROM directory
Covers over 3,500 CD-ROM and multimedia CD titles
worldwide. Published twice a year by Task Force Pro Libra.

Gale directory of databases
A two volume directory, Volume 2 of which has entries for over
3,500 CD-ROM and related products worldwide. Published
twice a year by Gale Research Inc.

8.7 Libraries and other services providing business information

This section starts with descriptions of three major national
business services: the British Library's Business Information
Service (BIS), the Department of Trade and Industry's Export
Market Information Centre (EMIC), and the information services
offered by the Chartered Institute of Marketing (CIM). This is
followed by brief notes on public libraries and other business
information services, then an evaluation of the value of what is
currently available on the Internet.

8.7.1 The British Library Business Information Service (BIS)

The Business Information Service provides comprehensive access
to international business information and is the major business
library in the UK. The business collections are housed on the
lower ground floor of the Holborn Reading Room of the Science
Reference and Information Service and any member of the public
can visit the library. There is an enquiry desk within the business
reading room where specialist information officers deal with

business enquiries and give visitors to the reading room any help they need to use the collection of reference works. The major categories of business information source held are:

- market research reports
- trade directories
- trade journals
- company annual reports, house journals and product literature
- business journals
- business CD-ROMs.

Many of these resources are described in more detail in the section above on published sources of information. For those unable to visit the Library to carry out their own research, the Business Information Service offers two enquiry services.

8.7.2   The British Library Lloyds Bank Business Line

This is a telephone information service operated by experienced staff of the Business Information Service. It is partly funded by sponsorship from Lloyds Bank PLC and is offered free, except for the cost of the call. Because of the popularity of this service help has to be limited to 10 minutes per call, but there is no limit to the range of information offered; typical enquiries which can be answered are a company's phone number, the name of its managing director and its area of activity, or details of suppliers of a particular product. Tel. 0171 412 7454/7977.

8.7.3   The Business Information Research Service

This is a fee-based service which offers in-depth research on all areas of business information. It can provide market overviews, company profiles, mailing lists and much more. In addition to the British Library collections, researchers providing this service also have access to an international range of online databases. Tel. 0171 412 7457, fax 0171 412 7453.

8.7.4 Export Market Information Centre (EMIC)

Run by the Department of Trade and Industry (DTI), this Centre has the UK's most comprehensive range of statistical sources on worldwide trade and production, and other key economic statistics relevant to marketing decisions. Its collection of other material of interest to exporters includes mail order catalogues, which could help the inventor investigate the competition and gaps in a market, as well as giving examples of presentation and pricing. A range of statistical, contact and marketing information is also available on CD-ROM.

EMIC also provides fax access to the Export Intelligence database, INFAX, and there is a bookshop with a full range of DTI publications relating to exporting.

Exporters, market researchers, consultants and business people are able to use EMIC free of charge and without prior appointment. EMIC's staff are able to signpost enquirers to other DTI services for exporters, such as export promoters and country experts. (See Appendix 1.)

8.7.5 The Chartered Institute of Marketing (CIM)

The Institute's information services are primarily to serve members of the Institute, but many of its resources are also available to non-members. The Institute produces a wide range of reading lists, and offers a priced information service. Those who are not members should contact the librarian to find out what is available, and about fees where applicable. (See Appendix 1.)

8.7.6 Public library services

Public library provision of business information varies greatly with size and subject specialisation of the library, but the current encouragement of business enterprise has led to an increase in specialist information services. The reader should refer to Section 7.5.3 on public libraries; although the focus there is on technical information, all the general comments apply equally to business information. The specific libraries named in that section as major sources of technical information are also the best source in their area for business information. Most have staff trained to deal with business enquiries, and many have specialist business collections. The Scottish Science Library, for instance, has the Scottish Business Information Service, which is a major resource offering all the kinds of information mentioned above in the description of the British Library's Business Information Service.

Several of the Patents Information Network libraries have their patents information and technical collections integrated with a specialist business service, as at Coventry's Lanchester Technical Information Service, set up to serve Small and Medium-sized Enterprises (SMEs). Others offer a separate business library, or services like Liverpool's Small Business Information Unit.

Two of the London public library services are known for their specialist business collections, but ask readers to note that, like other public libraries, their priority is to serve the local community. The Corporation of London City Business Library is specifically intended for those who live, work or study within the "square mile" of the City. Similarly, while the Business Library of the Westminster Reference Library has a large collection of business material which any visitor may consult, their enquiry service is intended for local people. Those from outside the area should approach their own libraries, or one of the regional or national libraries which provide help on a wider basis.

8.7.7 Other business information services

Governments wish to encourage innovative enterprise, but until recently anyone thinking of starting up in business or expanding a small business faced a bewildering range of advice and information from the different local support agencies. Now it has been made easier to identify and access the assistance which best suits their needs, with single-point entry telephone helplines in England, Scotland, Wales and Northern Ireland. The services are described briefly below.

England
Business Link Signpost Line: tel. 0345 567765.

Ringing this local rate number will put you in touch with your nearest Business Link office. Business Links are a network of independent local business information and advice centres, run by partnerships which include Chambers of Commerce, Local Authorities, Training and Enterprise Councils (TECs), Local Enterprise Agencies (LEAs), and the Department of Trade and Industry. There are over 200 Business Links, but where there is not one yet established in an area the Business Link Signpost Line will put you in touch with the most appropriate local organisation, probably a TEC or LEA. Contact details of Business Links may also be found through their Internet site, the address for which is <http://www.businesslink.co.uk/>.

While intended primarily to improve the competitiveness of small and medium enterprises, Business Links offer help and advice on a wide range of business topics, both on a personal basis and through events and courses. All have Personal Business Advisers, and many provide access to Innovation and Technology Counsellors. After the initial advice sessions users of Business Links may be referred to priced consultancy services for further specialist help. At the time of writing the DTI is funding an Innovation Credit scheme which provides support for those investigating the potential of an invention, and it is accessed through the Business Links.

A European initiative to establish a network of business advice centres to encourage innovation has resulted in a network of offices in England, known as Business Innovation Centres, or BICs. Some operate under their own local name, but a list of those currently in operation is given in Appendix 1, under the heading Business Innovation Centres. The offices listed under another European programme, Innovation Relay Centres, listed in Appendix 1, can also help with business information.

Scotland

Scottish Business Shop Network: tel. 0800 787878.

Ringing this freephone number will connect you to the nearest local Business Shop in Scotland. Business Shops offer access to local, national and international business information and provide a service to established and potential enterprises. In all parts of

Scotland, except the Highlands and Islands, The Business Shop Network is supported by Scottish Enterprise, Local Enterprise Companies and other partner organisations. The efforts and central funding from Scottish Enterprise, the regional economic development agency for most of Scotland, has led to a variety of help for the innovative entrepreneur. There are technology and economic development units, with varying names and sponsors, in every region, and Entrepreneurship Centres in the universities. The Business Shops and the LECs are the pointers to all these initiatives.

In the Highlands and Islands information and development assistance is delivered through LECs, which are contracted by Highlands and Islands Enterprise. The first point of contact for inventors should be the LECs listed in Appendix 1.

Wales

Business Connect Wales helpline: tel. 0345 969798.

Ringing this local rate number will automatically connect you to the nearest Business Connect office. Business Connect is an advice and support network with links to dozens of specialist services. Each Business Connect is supported by the Welsh Office and a consortia of other government and local organisations.

Help for small businesses and independent innovators is also available from the Welsh Business Innovation Centres (BICs), Innovation Wales in the south and Snowdonia BIC in the north. These are "not for profit" organisations, part of the European Business Network, and offer a free initial assessment of an idea, followed by advice on all aspects of developing it further.

Northern Ireland

The first contact point for business information and help is LEDU, The Small Business Agency for Northern Ireland, tel. 01232 491031, or e-mail ledu@nics.gov.uk. LEDU offers a wide range of services to those starting up in business and those already in business, including access to a network of Business Counsellors, who give initial advice and evaluation free of charge. The six Regional Offices have business start-up publications, and run the Business Start Programme of assistance, and their contact details can be found in Appendix 1, under the LEDU heading. Local LEDU offices can also arrange appointments with its Product Development Workshop, which is equipped to help design and make prototypes of new products.

LEDU is also the host organisation of a European Information Centre (EIC) and the Innovation Relay Centre which is based at LEDU headquarters, and offers information for those in business. Most enquiries can be answered free of charge.

Local Enterprise Agencies (LEAs) also provide advice and training for those new to business, and can help with business premises and loans. Details of the nearest LEA will be provided by LEDU, at the telephone number above.

8.7.8   Business information on the Internet

The provision of business information is probably the fastest-changing aspect of the Internet service at the moment, so the most important advice is to check what is currently available before setting up an Internet connection primarily as a source of business information. Ways of doing this include reading recent issues of the Internet periodicals, and reviews and discussions in general business periodicals. It is increasingly easy to get access to the Internet at local libraries for a trial exploration of what is available, or at the growing number of shops or cafés which offer their customers pay-as-you-use sessions on machines connected to the Internet.

At the time of writing the Internet is not a major source of UK business information, and has not been much used by information providers because of the heavy bias to US material, the lack of quality control, and the limits to advertising material on the Internet. Two at least of these factors are changing: there is a noticeable increase in business information relevant to UK users which will build up to a useful resource in time, and the approach to the use of the Internet for marketing is changing. The latter aspect is of interest to inventors not only for promoting their own product, but as a source of market and product information to use at an early stage of their business planning. Discussion groups can be a source of information on products and services, and comments will usually be allowed if a legitimate contribution to the group's interest, but pure advertising is frowned on as it could swamp the discussion. This does not of course apply to the dedicated ads newsgroups, the Internet equivalent to classified ads columns in the press. World Wide Web sites on the Internet (see Section 7.6.4 for a brief introduction to this aspect of the Internet) can include any information on products or services, so you can search these for items of interest or hire your own Web site to publicise what you have to offer. New ways of hiring advertising space on the Internet are being developed, and should be checked at the time of reading.

The Department of Information Science, University of Strathclyde, maintains an index to business sources on the Internet. The Internet address is <http://www.dis.strath.ac.uk/business/index.html>. This is a selective guide, organised by types of information, for example company profiles and financial information, country information, and is updated regularly. The most common type of UK business data on the Internet at the moment is directory information, with the coverage increasing daily and normally offering better currency than sources of information in other media.

To sum up: at the moment the traditional sources still provide much better organised access to most types of business information, and for anyone with ready access to a large business library it is not worth setting up an Internet connection solely for that purpose. As more UK business information is made available this may change and already it can be a useful source for anyone with access. For anyone considering setting up a personal connection there is a section at the end of Chapter 7 giving basic information, and a list of some service providers is given in Appendix 6.

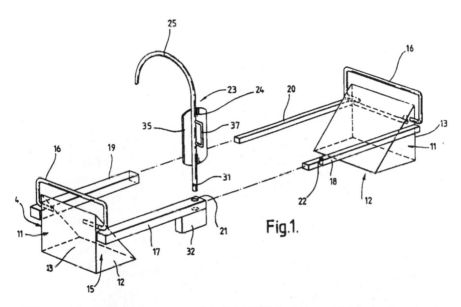

Fig.1.

Figure 12. British patent number 2112725. Lionweld Limited's wheel clamp patent, applied for in 1981.

Appendix 1
Organisations

Please note that the institutions, companies and other bodies listed in this appendix are given as examples of sources of more detailed help and information. This list is indicative rather than comprehensive, and details of other bodies offering help in the same fields can be found in the publications listed in the bibliography or in other directories. The inclusion of names in the list below should not be taken to indicate an endorsement of the services offered.

Advanced Technology Centre
University of Warwick
Coventry CV4 7AL
Tel: 01203 524722/524873
Fax: 01203 524878
Website: http://www.warwick.ac.uk/atc/
The Rapid Prototyping and Tooling Centre advises companies on the most effective ways of applying rapid prototyping and the related technologies.

Barclays Bank PLC
Barclays Information Centre
Freepost CV2462
Stratford upon Avon
Warwickshire CV37 9BR
Tel: Barclays Business Line: 0800 400170
Fax: 01789 292341
Website: http://www.barclays.co.uk/
Publishes *Barclays Small Business Guide*. Local branches also have free booklets of advice for starting up in business.

British Chambers of Commerce
Manning House
22 Carlisle Place
London SW1P 1JA
Tel: 0171 565 2000
Fax: 0171 565 2049
Provides enquirers with the address of their local Chamber of Commerce.

British Venture Capital Association
Essex House
12-13 Essex Street
London WC2R 3AA
Tel: 0171 240 3846
Fax: 0171 240 3849
Email: bvca@bvca.co.uk
Website: http://www.brainstorm.co.uk/BVCA
Publishes free information on sources of UK venture capital.

Brunel Innovation
Design Research Centre
Brunel University
Egham
Surrey TW20 0JZ
Tel: 01784 431341 ext 388
Fax: 01784 472879
Website: http://http1/brunel.ac.uk:8080/faculty/tech/des/
           research/research.html
The Centre has its own facilities for modelmaking and prototyping, and access to University-wide expertise. Its Inventor Programme can be accessed via referral from the Business Links, and after a commercial viability study can offer help with developing new product ideas, including finding appropriate partners.

Business Connect Wales helpline
Ringing 0345 969798 will automatically connect you to the nearest office of the Business Connect advice and support network.

Business Innovation Centres (BICs)
These are members of the European Business Network (EBN).
They offer a range of help to SMEs and individuals.

Bangor
Snowdonia BIC
Llys y Fedwen
Parc Menai
Bangor
Gwynedd LL57 4BF
Tel: 01248 671101
Fax: 01248 671102

Barnsley
Barnsley BIC Ltd
Innovation Way
Barnsley
South Yorkshire S75 1JS
Tel: 01226 249590

Birmingham
Birmingham Technology Ltd
Love Lane
Aston Triangle
Birmingham B7 4BJ
Tel: 0121 359 0981
Fax: 0121 359 0433

Bradford
WYEBIC
2nd Floor
Mercury House
4 Manchester Road
Bradford BD5 0QL
Tel: 01274 841300
Fax: 01274 841302

Cambridge
St John's Innovation Centre
St John's Park
Cowley Road
Cambridge CB4 4WS
Tel: 01223 421194
Fax: 01223 420844

Cardiff
Innovation Wales CBTC
Senghennydd Road
Cardiff CF2 4AY
Tel: 01222 667041
Fax: 01222 373436

Enfield
Lee Valley BIC
Unit 4
The Arena
Mollison Avenue
Enfield
Middlesex EN3 7NL
Tel: 0181 805 8100
Fax: 0181 805 8101

Londonderry
NORIBIC Ltd
Ulster Science and Technology Park
Bumcrana Road
Londonderry
BT48 0NB
Tel: 01504 264242
Fax: 01504 269025

Manchester
Greater Manchester BIC
Windmill Lane
Denton
Manchester M34 3QS
Tel: 0161 337 8648
Fax: 0161 337 8651

Preston
Lancashire Enterprises plc
Enterprise House
17 Ribblesdale Place
Preston PR1 3NA
Tel: 01772 203020
Fax: 01772 887110

Stafford
Staffordshire & Black Country BIC
Staffordshire Technology Park
Beaconside
Stafford ST18 0AR
Tel: 01785 226598
Fax: 01785 220302

Sunderland
Sunderland BIC
Sunderland Enterprise Park
Riverside
Sunderland SR5 2TA
Tel: 0191 516 0160
Fax: 0191 516 8159

Wrexham
NEWTECH Innovation Centre
Croesnewydd Hall
Wrexham
LL13 7YP
Tel: 01978 290694
Fax: 01978 290705

York
York Innovation Ltd
York Enterprise Centre
1 Davygate
York YO1 2QE
Tel: 01904 679677
Fax: 01904 625287

Business Link Signpost line (England)
Ringing 0345 567765 will put you in touch with the nearest office
of this network of business information and advice centres. Section
8.7.7 gives more details.

Business Shops (Scotland)
Ring the freephone number 0800 787878, for connection to your
nearest Business Shop or Local Enterprise Company.

Chartered Institute of Marketing
Moor Hall
Cookham
Berkshire SL6 9QH
Tel: 01628 427500
Fax: 01628 427499
Email   marketing@cim.co.uk
Website: http://www.cim.co.uk
Provides a comprehensive consultancy service, either on a direct
project managed basis or by matching clients to consultants on its
register. Other services include the provision of marketing
information through CIM's Information and Library Service, and
mail order publications via CIM Direct.

Chartered Institute of Patent Agents
Staple Inn Buildings
High Holborn
London WC1V 7PZ
Tel: 0171 405 9450
Fax: 0171 430 0471
Website: http://www.users.dircon.co.uk/~cipa/

Maintains a regional directory of firms of patent agents. Publishes a free information pack on protecting all kinds of intellectual property.

European Venture Capital Association (EVCA)
Keiberpark-Minervastraat 6
Box 6
B-1930 Zaventem
Belgium
Tel: 00 32 2 7150020
Fax: 00 32 2 7250704
Represents over 200 leading venture capital companies in Europe. Produces directory with members' details, and information on funds and investment criteria, available for reference in some libraries and information centres.

Export Market Information Centre (EMIC)
Kingsgate House
66-74 Victoria Street
London SW1E 6SW
Tel: 0171 215 5444/5445
Fax: 0171 215 4231
Email: EMIC@as001.ots.dti.gov.uk
Run by the Department of Trade and Industry. Section 8.7.4 gives details of the service.

Hatrics
The Southern Information Network
81 North Walls
Winchester SO23 8BY
Tel: 01962 846115
Fax: 01962 856615
Email: libseb@hants.gov.uk
Website: http://www.soton.ac.uk/~hatrics/
A subscription-based information network of commercial and industrial companies, public libraries, academic institutions, government organisations and research establishments.

Innovation Relay Centres (IRCs)
These are some of the advisory centres financed by the European Union to encourage participation in its research and technological development activities. Some are lead offices for others in the area. *Website:* http://www.cordis.lu/cordis/irc.html

Belfast
LEDU
LEDU House
Upper Galwally
Belfast BT8 4TB
Tel: 01232 491031
Fax: 01232 691432
Website: http://alexandra.nics.gov.uk/ledu/irc/

Cambridge
The Technology Broker Ltd
Station Road
Longstanton
Cambridge CB4 5DS
Tel: 01954 261199
Fax: 01954 260291

Cardiff
Welsh Development Agency (WDA)
Principality House
The Friary
Cardiff CF1 4AA
Tel: 01222 828739
Fax: 01222 640030

Coventry
Coventry University Enterprises Ltd (CUE)
Priory Street
Coventry CV1 5FB
Tel: 01203 838140
Fax: 01203 221396

Farnborough
Defence & Evaluation Research Agency (DERA)
Q108 Building
Farnborough GU14 6TD
Tel:　01252 392343
Fax:　01252 393318

Glasgow
Euro Info Centre Ltd (EIC)
21 Bothwell Street
Glasgow G2 6NL
Tel:　0141 221 0999
Fax:　0141 221 6539

Sunderland
RTC North Ltd
3D Hylton Park
Wessington Way
Sunderland SR5 3NR
Tel:　01915 498299
Fax:　01915 489313

Innovation and Technology Management
Technology Transfer
Lancashire Enterprises plc
17 Ribblesdale Place
Winckley Square
Preston PR1 3NA
Tel:　01772 203020
Fax:　01772 887110
Email:　itm@lancs-ent.co.uk
Website: http://www.worldserver.pipex.com/aim/AIM.htmld/
　　　　　LCE.L/index.html
Helps small companies develop and market new products and technologies. Offers a comprehensive package of services tailored to the needs of individual companies to help them reap the benefits of innovation and technology.

Institute of Chartered Accountants in England & Wales
Chartered Accountants Hall
PO Box 433
Moorgate Place
London EC2P 2BJ
Tel: 0171 920 8680
Fax: 0171 920 8621
Website: http://www.icaew.co.uk/
Will refer enquirers to their local district administrator who has
information on members with appropriate experience.

Institute of Chartered Accountants of Scotland
27 Queen Street
Edinburgh
Midlothian EH2 1LA
Tel: 0131 225 5673
Fax: 0131 225 3813
The Member Services Department can provide a free list of local
Chartered Accountants.

The Institute of International Licensing Practitioners Ltd
Suite 73
Kent House
87 Regent Street
London W1R 7HF
Tel: 0171 287 0200
Fax: 0171 287 0400
Will circulate members with details of products available for
licensing (for a small fee).

The Institute of Patentees and Inventors
Suite 505A
Triumph House
189 Regent Street
London W1R 7WF
Tel & Fax: 0171 242 7812
Website: http://www.invent.org.uk/
Membership entitles inventors to advice and reduced–cost access
to information services.

Institute of Trade Mark Agents
Canterbury House
2-6 Sydenham Road
Croydon
Surrey CR0 9XE
Tel: 0181 686 2052
Fax: 0181 680 5723
Website: http://itma.org.uk/
Publishes a geographical list of members, and produces a short guide for those considering registering a trade mark.

Institution of Electrical Engineers
Savoy Place
London WC2R 0BL
Information Services (*see* Appendix 2 for Library Services):
Tel: 0171 344 8428/9
Fax: 0171 497 3557
Email: bite@iee.org.uk
Website: http://ampere.iee.org.uk/
Offers charged searching services covering technical, market, company and legislative information.

Intellectual Property Consultancy
42 Saint George's Road
London NW11 0LR
Tel: 0181 458 5358
Fax: 0181 458 0543
Email: jeremy_phillips@link.org
Jeremy Phillips writes and lectures internationally in many fields of intellectual property. He is currently intellectual property consultant to London-based solicitors Slaughter and May, editor of the *European trade mark reports*, the *Butterworths intellectual property law handbook* and the *Aslib guide to copyright*, as well as consultant editor of *Managing Intellectual Property*.

The Intellectual Property Development Confederation

72a Bedford Place
Southampton SO15 2DS
Tel: 01703 570101
Fax: 01703 570102
Membership organisation which offers access to a wide variety of information, support services and contacts, and publishes the *Inventors World* magazine.

Inventorlink Products Ltd

5 Clipstone Street
London W1P 7EB
Tel: 0171 323 4323
Fax: 0171 323 0286
Email: inventorlink@dial.pipex.com
Website: http://london.globalnews.com/inventorlink/
Offers specialist help in launching new products, published in the monthly illustrated *Inventions International* bulletin. Has research, marketing, PR and licensing departments, and negotiates royalty agreements, sales, joint ventures, and new company formations.

Kent Technology Transfer Centre (KTTC)

Research & Development Building
University of Kent
Canterbury
Kent CT2 7PD
Tel: 01227 763414
Fax: 01227 763424
Offers a range of services to companies in the south-east to promote use of new technology.

The Law Society of Scotland

26 Drumsheugh Gardens
Edinburgh EH3 7YR
Tel: 0131 226 7411
Fax: 0131 225 2934
Provides enquirers with the names of solicitors in their locality who deal with intellectual property.

LEDU
LEDU House
Upper Galwally
Belfast BT8 4TB
Tel: 01232 491031
Website: http://www.nics.gov.uk/ledu/leduhome.htm
This office is the LEDU headquarters, and also hosts a European
Business Information Centre.

LEDU Regional Offices
These are the LEDU regional offices operated by LEDU, The
Small Business Agency for Northern Ireland. Section 8.7.7 gives
more details.

LEDU Belfast Regional Office
25–27 Franklin Street
Belfast BT2 8DT
Tel: 01232 242582

LEDU North Eastern Office
Clarence House
86 Mill Street
Ballymena BT43 5AF
Tel: 01266 49215

LEDU North Western Office
13 Shipquay Street
Londonderry BT48 6DJ
Tel: 01504 267257

LEDU South Eastern Office
43–45 Frances Street
Newtownards BT23 3DX
Tel: 01247 813880

LEDU Southern Office
6–7 The Mall
Newry BT34 1BX
Tel: 01693 62955

LEDU Western Office
Kevlin Buildings
47 Kevlin Avenue
Omagh BT78 1ER
Tel: 01662 245763

Local Enterprise Companies in Scotland

Argyll and the Islands Enterprise
The Enterprise Centre
Kilmory Industrial Estate
Lochgilphead PA31 8SH
Tel: 01546 602281
Fax: 01546 603964

Caithness and Sutherland Enterprise
Scapa House
Castlegreen Road
Thurso
Caithness KW14 7LS
Tel: 01847 896115
Fax: 01847 893383

Dumbartonshire Enterprise
2nd Floor
Spectrum House
Clydebank
Glasgow G81 2DR
Tel: 0141 951 2121

Dumfries and Galloway Enterprise
Solway House
Dumfries Enterprise Park
Tinwald Downs Road
Dumfries DG1 3SJ
Tel: 01387 245000
Fax: 01387 246224

Enterprise Ayrshire
17–19 Hill Street
Kilmarnock KA3 1HA
Tel: 01563 26623

Fife Enterprise
Kingdom House
Saltire Centre
Glenrothes
Fife KY6 2AQ
Tel: 01592 623000
Fax: 01592 623149

Forth Valley Enterprise
Laurel House
Laurelhill Business Park
Stirling FK7 9JQ
Tel: 01786 451919
Fax: 01786 478123

Glasgow Development Agency
Atrium Court
50 Waterloo Street
Glasgow G2 6HQ
Tel: 0141 204 1111

Grampian Enterprise Ltd
27 Albyn Place
Aberdeen AB1 1YL
Tel: 01224 211500

Inverness and Nairn Enterprise
Castle Wynd
Inverness IV2 3DW
Tel: 01463 713504
Fax: 01463 712002

Lanarkshire Development Agency
New Lanarkshire House
Strathclyde Business Park
Bellshill
Motherwell ML4 3AD
Tel: 01698 745454

Lochaber Ltd
St Mary's House
Gordon Square
Fort William PH33 6DY
Tel: 01397 704326
Fax: 01397 705309

Lothian & Edinburgh Enterprise Ltd
Apex House
99 Haymarket Terrace
Edinburgh EH12 5HD
Tel: 0131 313 4000

Moray, Badenoch and Strathspey Enterprise
Elgin Business Centre
Maisondieu Road
Elgin IV30 1RH
Tel: 01343 550567
Fax: 01343 550678

Orkney Enterprise
14 Queen Street
Kirkwall KW15 1JE
Tel: 01856 874638
Fax: 01856 704130

Renfrewshire Enterprise
27 Causeyside Street
Paisley PA1 1UL
Tel: 0141 848 0101

Ross and Cromarty Enterprise
High Street
Invergordon IV18 0DH
Tel: 01349 853666
Fax: 01349 853833

Scottish Borders Enterprise
Bridge Street
Galashiels TD1 1SW
Tel: 01896 758991
Fax: 01896 758625

Scottish Enterprise Tayside
Enterprise House
45 Lindsay Street
Dundee DD1 1HT
Tel: 01382 23100

Shetland Enterprise
Toll Clock Shopping Centre
26 North Road
Lerwick ZE1 0PE
Tel: 01595 693177
Fax: 01595 693208

Skye and Lochalsh Enterprise
King's House
The Green
Portree
Isle of Skye IV51 9BS
Tel: 01478 612841
Fax: 01478 612164

Western Isles Enterprise
3 Harbour View
Cromwell Street Quay
Stornoway
Isle of Lewis HS1 2DE
Tel: 01851 703625

London Enterprise Agency (LEntA)
4 Snow Hill
London EC1A 2BS
Tel: 0171 236 3000
Fax: 0171 329 0226
Email: lenta@itl.net
Enterprise agency funded by 19 of Britain's largest companies and the Corporation of London. Has a specialist consultancy and a range of products to help develop a business.

Licensing Executives Society Britain and Ireland (LES)
c/o MEDTAP International
27 Gilbert Street
London W1Y 1RL
Tel: 0171 290 9600
Fax: 0171 629 9705
Website: http://www.personal.u-net.com/~patents/home.htm
Welcomes enquiries on exploitation of industrial property and all forms of technology transfer.

LINC Head Office
4 Snow Hill
London EC1A 2BS
Tel: 0171 236 3000
Fax: 0171 329 0226
LINC (Local Investment Networking Company) is a nationwide business introduction service, linking private investors (business angels) with businesses seeking equity investment between £10,000 and £250,000. LINC is sponsored by Barclays, Midland, NatWest, The Royal Bank of Scotland, and accountants Kingston Smith.

Lloyds Bank Plc
Victoria House
Southampton Row
London WC1B 5HR
Tel: 0345 309781
Fax: 0171 404 2073

These telephone and fax numbers connect you to the Bank's Business Centre, which offers free advice to inventors thinking of starting up in business.

MAREL Trinity Enterprise Centre
Furness Business Park
Barrow-in-Furness
Cumbria LA14 2PN
Tel: 01229 820611
Fax: 01229 820438
Aims to take inventions through to the market, using the engineering resources of the Furness region of the North West of England. Will consider projects of all types and provides free preliminary assessments of them.

Marketing Aids (UK) Ltd
Park House
10 Ray Park Avenue
Maidenhead
Berkshire SL6 8DS
Tel: 01628 36679
Fax: 01628 580468
Marketing Aids (UK) Ltd provides consultancy services in the management and exploitation of intellectual property, including the establishment of new businesses. Clients include academic institutions, companies both large and small, and private inventors. Roy Fuscone, Director of the company, is a marketer and an inventor.

Materials Information Service
Institute of Materials
1 Carlton House Terrace
London SW1Y 5DB
Tel: 0171 839 4071
Fax: 0171 839 5513
This enquiry line will bring you information and advice on all aspects of engineering materials and is backed up by local advisers in England, Scotland, and Wales, who can visit enquirers and will sign confidentiality agreements.

Merseyside Innovation Centre
131 Mount Pleasant
Liverpool L3 5TF
Tel: 0151 708 0123
Fax: 0151 707 0230
Website: http://www.connect.org.uk/merseyworld/MIC/
Provides technical, advisory and consultancy support on evaluation
of ideas, product development, prototyping, feasibility studies,
marketing and funding.

Midland Bank plc
Customer Information Service
PO Box 757
Hemel Hempstead
Hertfordshire HP2 4SS
Website: http://www.midlandbank.co.uk/bb/mdbkusb.htm
Tel: 0345 180180
Phone this number for free help for those starting in business, or
alternatively contact your local branch.

NatWest UK
Innovation & Growth Unit
Level 10
Drapers Gardens
12 Throgmorton Avenue
London EC2N 2DL
Tel: 0171 454 2847
Fax: 0171 454 2610
The Unit can direct you to the nearest branch with a specialist
innovation adviser.

NEL
The Scottish Technology Park
East Kilbride
Strathclyde G75 0QU
Tel: 013552 20222
Fax: 013552 272999
Website: http://www.neag.co.uk/nel.htm

Physical Properties Data Service: *Tel:* 01355 272527; other information services include Materials and Engineering Consultancy Service and the National Agency for Finite Elements and Standards.

NIMTECH
Alexandra House
Borough Road
St Helens WA10 3TN
Tel: 01744 453366
Fax: 01744 453377
Website: http://www.merseyworld.com/nimtech
Network of innovative organisations in the north-west offering technical support and contacts to companies and individuals. Has the DEVICE programme to assist innovators all over the UK to get their ideas to the marketplace.

NIS Invotec Ltd
5 Chorley West Business Park
Ackhurst Road
Chorley
Lancashire PR7 1NL
Tel: 01257 266707
Fax: 01257 266687
Provides services for new manufacturing.

The Northern Ireland Technology Centre
The Queen's University of Belfast
Cloreen Park
Malone Road
Belfast BT9 5HN
Tel: 01232 664393
Fax: 01232 663715
Website: http://www.nitc.qub.ac.uk/
Advises on the design and engineering aspects of new product development.

The Patent Office
Cardiff Road
Newport
Gwent NP9 1RH
Website: http://www.patent.gov.uk/
Enquiry line: 0645 500505
This local rate number can be used to request the Patent Office's free information pack covering patents, trade marks, designs and copyright, or to be put in touch with the appropriate person to answer individual queries.

Patent Plan
Head Office
Unit M1
Hilton Business Park
East Wittering
Chichester PO20 8RL
Tel:　　01243 671885
Fax:　　01243 671884
A specialist patent drafting service enabling the filing of patent applications at minimum cost. The service is specifically designed to meet the needs of small companies and inventors. Patent Plan offers a nationwide service.

Product Development Tayside Ltd
Bell Street
Dundee DD1 1HG
Tel:　　01382 308199
Fax:　　01382 308290
Organisation offering practical support and engineering input to individuals and small companies who have ideas for new products identified as being commercially viable.

Royal Society of Chemistry (RSC)
Library & Information Centre (LIC)
Burlington House
Piccadilly
London W1V 0BN

Tel: 0171 437 8656
Fax: 0171 287 9798
Email: library@rsc.org
Website: http://chemistry.rsc.org/rsc/welcome.htm

The LIC offers a priced technical and business enquiry service (which includes online searching) on all aspects of chemical information. The RSC produces a wide range of publications (journals, newsletters and books) on chemistry and related subjects.

Science Line
Tel: 0345 600 444

Managed by Broadcasting Support Services; funded by the Wellcome Trust, the Royal Society, the Research Councils and the DTI to provide members of the public with a free telephone enquiry service on any scientific topic. Calls are at the local rate, and lines are open 1.00–7.00 every weekday.

Scottish Enterprise
21 Bothwell Street
Glasgow G2 6NL
Tel: 0141 221 6711
Fax: 0141 221 6722
Website: http://www.scotent.co.uk/

Has an Expert Help Programme, in partnership with the Local Enterprise Companies, to assist SMEs with innovation and other business needs.

SUPERNET
Tel: 0645 121100

A business service which helps solve technological problems. Appropriate organisations able to clarify the problem will be identified and a named contact given. Initial costs are covered by DTI, further commercial work is negotiated on an individual basis. The usual way to access SUPERNET help is through the Business Link, Business Connect, or Business Shops networks, but it can be accessed direct if necessary.

TEChINVEST
South & East Cheshire TEC Ltd
PO Box 37
Dalton Way
Middlewich
Cheshire CW10 0HU
Tel: 01606 734308/734288/737009
Fax: 01606 734201
Links companies to sources of funding, particularly private investors. Companies in the North-West get priority, but the service is open to all.

TIE (UK) Ltd
6 Aztec Row
Berner's Road
London N1 0PW
Tel: 0171 704 9702
Fax: 0171 704 9594
Specialises in technology transfer on a worldwide basis, covering both the licensing in and out of innovative technology as well as providing advice on seeking seed and venture capital and other appropriate means of financing.

Venture Capital Report
The Magdalen Centre
Oxford Science Park
Oxford OX4 4GA
Tel: 01865 784411
Fax: 01865 784412
Website: http://www.demon.co.uk/vcr1978/
Produces a monthly report to introduce business angel investors to businesses with innovative products.

The Welsh Design Advisory Service
Design Engineering Research Centre
University of Wales Institute
Cardiff Western Avenue
Cardiff CF5 2YB

Tel: 01222 506720
Fax: 01222 506970

Funded by the Welsh office; offers free advice on all aspects of design to SMEs in Wales.

WEMTECH
Bordesley Hall
The Holloway
Alvechurch
Birmingham B48 7QQ
Tel: 015275 95066
Fax: 015275 95033

World Metal Index
Surrey Street
Sheffield S1 1XZ
Tel: 0114 273 4714/4744
Fax: 0114 273 5009

National resource centre which provides information on standards, suppliers, and technical literature. Charges are by subscription or for service provided.

Yorkshire and Humberside RTN Ltd
Batley Business and Technology Centre
Technology Drive
Batley
West Yorkshire WF17 6ER
Tel: 01924 423430
Fax: 01924 445059
Website: http://www.dbms.co.uk/bbtc/

Points companies and individuals to the most appropriate business support agency and technology provider. Innovation Relay Centre for the North-East.

Appendix 2
Libraries

This list of libraries open to the general public starts with the national libraries, and then lists each library in the Patents Information Network under its town. Contact details for other libraries are available from your nearest reference library, in the directories described in Chapter 7 (and in Appendix 8).

National libraries

The British Library Science Reference and Information Service
25 Southampton Buildings
London WC2A 1AW
Science and technology enquiries
    Tel: 0171 412 7494
    Fax: 0171 412 7495
Patents enquiries
    Tel: 0171 412 7919/7920
    Fax: 0171 412 7480
Business Information Service enquiries
    Tel: 0171 412 7457
Open 9.30–9.00 on weekdays; 10.00–1.00 on Saturdays; Foreign Patents Annexe open 9.30–5.30 Monday–Friday
(Life sciences are located in the Aldwych Reading Room, 9 Kean Street, London WC2B 4AT; Tel: 0171 412 7288; open 9.30–5.30 Monday to Friday)
More details in Section 7.5.1.

The British Library Document Supply Centre
Boston Spa
Wetherby
West Yorkshire LS23 7BQ
Tel: Customer Services: 01937 546060
More details in Section 7.5.2.

Scottish Science Library
National Library of Scotland
33 Salisbury Place
Edinburgh EH9 1SL
Tel: 0131 226 4531
Fax: 0131 220 6662
Scottish Business Information Service
     Tel: Helpdesk: 0131 667 9554
Open Monday, Tuesday, Thursday and Friday 9.30–5.00;
Wednesday 10.00–8.30; closed on Saturday.
More details in Section 7.5.3.

Llyfrgell Genedlaethol Cymru/The National Library of Wales
Aberystwyth
Dyfed SY23 3BU
Tel: 01970 623816
Fax: 01970 615709
Open weekdays 9.30–6.00; Saturday 9.30–5.00.
Apply to Secretary for reader's ticket; day tickets issued on proof of identity.

Other libraries open to the public now follow. Those which are members of the Patents Information Network are marked with an asterisk, and have collections of technical literature and business information.

Aberdeen

Business & Technical Department★
Central Library
Rosemount Viaduct
Aberdeen AB9 1GU
Tel: 01224 634622
Fax: 01224 636811
Open weekdays 9.00-8.00; Saturday 9.00-5.00.

Belfast

Science Library★
Central Library
Belfast Public Libraries
Royal Avenue
Belfast BT1 1EA
Tel: 01232 243233
Fax: 01232 332819
Open Monday and Thursday 9.30-8.00; Tuesday, Wednesday and
Friday 9.30-5.00; Saturday 9.30-1.00.

Queen's University of Belfast
Science Library
Chlorine Gardens
Belfast BT9 5EQ
Tel: 01232 245133 x4309
Fax: 01232 382636
Open weekdays 9.00-10.00; Saturday 9.00-12.30.
Non-students may use the library for reference; associate
membership, with borrowing rights, is available for a fee.

Birmingham

Patents Department★
Central Library
Chamberlain Square
Birmingham B3 3HQ

Tel: 0121 235 4537
Fax: 0121 233 4458
Open weekdays 9.00-8.00; Saturday 9.00-5.00

Bristol

Library of Commerce and Industry★
Central Library
College Green
Bristol BS1 5TL
Tel: 0117 929 9148
Fax: 0117 922 6775
Patents
     Tel: Patents department: 0117 929 9148
Open Monday to Thursday 10.00-7.30; Friday 9.30-7.30;
Saturday 9.00-5.00.

Cardiff

Central Library
St David's Link
Frederick Street
Cardiff CF1 4DT
Tel: 01222 382116
Fax: 01222 238642
Open Monday, Tuesday and Friday 9.00-6.00; Wednesday and
Thursday 9.00-8.00; Saturday 9.00-5.30.

Coventry

Lanchester Library★
University of Coventry
Much Park Street
Coventry CV1 2HF
Tel: 01203 838292
Fax: 01203 838686
Ring for details of general library opening hours, which vary.
Patents
     Tel: 01203 838167

Dundee

Commerce & Technology Department
Central Library
The Wellgate
Dundee DD1 1DB
Tel: 01382 434318
Fax: 01382 434642
Open weekdays 9.30-7.00, except Wednesday, 10.00-7.00;
Saturday 9.30-5.00.
Although not a member of the Patents Information Network, this
library has patents material and staff who can help with enquiries.

Edinburgh

Central Library
George IV Bridge
Edinburgh EH1 1EG
Tel: 0131 225 5584
Fax: 0131 225 5584
Open: weekdays 9.00-8.30 (Reference till 9.00); Saturday
9.00-1.00.

Glasgow

Science and Technology Department*
Mitchell Library
North Street
Glasgow G3 7DN
Tel: 0141 247 2931
Fax: 0141 287 2912
Open weekdays 9.00-9.00; Saturday 9.00-5.00.
Patents
      Patents Collection, Business Information
      Tel: 0141 287 2903/2904/2905

Leeds

Central Library★
Municipal Buildings
Calverley Street
Leeds LS1 3AB
Tel: 0113 247 8265/6
Fax: 0113 247 8268
Open Monday and Wednesday 9.00–8.00; Tuesday, Thursday, Friday 9.30–5.30; Saturday 9.00–5.00.
Patents
        Patents Information Unit
        32 York Road
        Leeds LS9 8TD
        Tel: 0113 214 3347
        Fax: 0113 248 8735
        Open weekdays 9.00–5.00; Saturday 9.00–1.00.

Liverpool

Science and Technology Library★
Central Libraries
William Brown Street
Liverpool L3 8EW
Tel: 0151 225 5442/5440
Fax: 0151 207 1342
Small Business Information Unit
        Tel: 0151 298 1613
        Fax: 0151 207 3109
Open Monday to Thursday 9.00–7.30; Friday 9.00–5.00; Saturday 9.00–5.00.

London

British Library Science Reference and Information Service
See National libraries, above.

Corporation of London City Business Library
1 Brewers' Hall Garden
London EC2V 5BX
Tel: 0171 638 8215 (enquiries)
          0171 480 7638 (recorded information)
Fax: 0171 260 1847
Open weekdays 9.30-5.00; closed on Saturday.
The service is intended for those who live or work in the area, but
the open access reference collection is available to all visitors.

Export Market Information Centre (EMIC)
Department of Trade and Industry
Kingsgate House
66-74 Victoria Street
London SW1E 6SW
Tel: 0171 215 5444/5445
Fax: 0171 215 4231
More details in Section 8.7.4.

Institution of Electrical Engineers
Savoy Place
London WC2R 0BL
Tel: 0171 344 5461
Library (*see* Appendix 1 for Information services)
        Fax: 0171 497 3557
        Email: libdesk@iee.org.uk
        Please note that borrowing from the Library is available to
        non-members only via interlibrary loan.
More details in Section 7.5.9.

Westminster Reference Library
35 St Martin's Street
London WC2H 7HP
Tel: 0171 641 2036
Open weekdays 10.00-7.00; closes at 5.00 on Saturday.
The service is intended for those who live or work in the borough,
but the reference collection is open to all.

Manchester

Technical Library★
Central Library
St Peter's Square
Manchester M2 5PD
Tel: 0161 234 1900
Fax: 0161 234 1963
Patents
　　Tel: 0161 234 1987
Open Monday to Thursday 10.00-8.00; Friday and Saturday 10.00-5.00.

Newcastle upon Tyne

Central Library★
Business and Science Library
Princess Square
Newcastle upon Tyne
Tel: 0191 261 0691
Fax: 0191 232 6893
Open Monday and Friday 9.30-8.00; Tuesday, Wednesday and Thursday 9.30-5.00; Saturday 9.00-5.00.
Patents
　　Tel: Patents Advice Centre: 0191 232 4601
　　Fax: 0191 232 4600
　　Usually open 9.30 -5.00 but check before visiting.

Plymouth

Central Library★
Reference Department
Drake Circus
Plymouth PL4 8AL
Tel: 01752 385906
Fax: 01752 385905
Open weekdays 9.00-7.30; closes at 4.00 on Saturday.

Portsmouth

Central Library★
Guildhall Square
Portsmouth PO1 2DX
Tel: 01705 819311
Fax: 01705 839855
Open weekdays 10.00–7.00; closes at 4.00 on Saturday.

Sheffield

Business and Technology Library★
Central Library
Surrey Street
Sheffield S1 1XZ
Tel: 0114 273 4711
Fax: 0114 273 5009
Patents
      Tel: 0114 273 4743
Open Monday 10.00–8.00; Tuesday, Thursday and Friday 10.00–5.30; Wednesday 1.00–8.00; Saturday 9.30–4.30.

Appendix 3
Intellectual Property Clinics

These clinics offer inventors an initial discussion of their ideas with a patent agent, who will advise on the next steps. The sessions are free, and it is always necessary to check in advance when a place is available as some clinics have long waiting lists.

Birmingham

Birmingham Central Library
Patents & Technology Centre
Chamberlain Square
Birmingham B3 3HQ
Tel: 0121 235 4537
Every Wednesday evening; appointment necessary.

Colchester

North & Mid Essex Chamber of Commerce
1/2 High Street
Colchester CO1 1DA
Tel: 01206 571187
Approximately every 2 months; appointment necessary.

Coventry

Lanchester Library
University of Coventry
Much Park Street
Coventry CV1 2HF
Tel: 01203 838508
First Tuesday of every month; appointment necessary.

Ipswich

Ipswich and Suffolk Chamber of Commerce
Suffolk Enterprise Centre
Russell Road
Ipswich IP1 2DE
Tel: 01473 210611 or 01206 571187
Approximately every 2 months; appointment necessary.

Liverpool

Science and Technology Library
Central Library
William Brown Street
Liverpool L3 8EW
Tel: 0151 225 5442
Alternate Tuesday evenings: appointment necessary.

London

Chartered Institute of Patent Agents
Staple Inn Buildings
London WC1V 7PZ
Tel: 0171 405 9450
Every Tuesday evening; appointment necessary.

Newcastle-upon Tyne

Patents Advice Centre
Central Library
Princess Square
Newcastle-upon Tyne NE99 1DX
Tel: 0191 232 4601
Thursdays, by prior arrangement.

Appendix 4
Patenting Costs

The costs of applying for the grant of a UK patent application vary greatly according to circumstances, so what follows is only an indication of probable minimum costs.

Every patent applicant must pay the standard Patent Office fees. A list of the current fees is available from the Patent Office (tel. 0645 500505). As at January 1997 a straightforward application, with no extra actions such as requests for amendment, entails the following payments:

Initial filing fee:	£25
Search fee:	£130
Examination fee:	£70
Annual renewal fees:	£110–£450 (from end of fourth year after application)

If patent protection is maintained for the full 20 years' life of the patent, the total minimum cost in Patent Office fees is £4095. Extra charges are made for work incurred by further actions such as amendments or restoration after late payment of the annual fees.

While it is possible for inventors to handle their own patent applications, it is almost always advisable to use the services of a qualified patent agent. The most important reason for doing so is that the protection given by the grant of a patent depends on the wording used to describe the invention, and the specification and claims are a legal language designed to be interpreted by lawyers in court. In addition to writing the patent application the patent agent may give professional help with various other aspects of developing the innovation, so the fees can vary widely.

As an example of the minimum costs to be expected, preparing the description and dealing with the formalities for the initial filing of a simple invention could cost around £800. The full cost of the further work involved in preparing a full specification and dealing with the Examiners' report could approach £1500, depending on the amount of work involved. These costs include the official Patent Office fees mentioned above.

The costs given above relate to obtaining a patent in the UK, and obtaining protection in other countries, if required, will give rise to fees and charges for each country.

When inventors approach a patent agent they should ask for details of the various stages of a patent application, and for an estimate of the total costs they are likely to incur.

The professional organisation for patent agents is the Chartered Institute of Patent Agents (CIPA), and it maintains a regional directory of firms of patent agents. Contact details of the Institute are given in Appendix 1.

Appendix 5
Sample Confidentiality Agreement

Supplied by Roy Fuscone

This is a sample confidentiality agreement between an inventor and a company involved in discussions concerning a patented product or process.

Letter of agreement re confidentiality

Subject: Your invention as disclosed in UK patent/patent application No.........

In discussion on the above subject it will be necessary for you (which shall include your representatives and employees) to inform this company (which shall include its representatives and employees) hereinafter referred to as XYZ, of confidential matters concerning the invention. This may include but not be limited to information covering the materials, design, process of manufacture, applications of the invention and other matters related thereto.

All information disclosed to this company by you relating to the invention and matters relating thereto will be kept confidential by XYZ. XYZ will also keep confidential any information relating to the invention or to matters relating thereto which may be acquired if XYZ is shown or loaned by you any documents, photographs, plant, machines, equipment, processes, specimens, products and the like.

In the event that our discussions do not lead to a business relationship between you and XYZ in connection with the invention, XYZ will return to you all copies of any information relating to the invention or to matters relating thereto where XYZ acquired such information from you, together with any photographs, plant, machines, equipment, specimens, products and the like which XYZ acquired from you. Such return shall be made within ten working days of the date of your request which shall be made by recorded delivery post or its equivalent to XYZ's address above. XYZ will retain no copies or other likenesses of any such information or items.

Any information required to be kept confidential by XYZ will not be disclosed by XYZ to any third party unless such a third party is required to know of such information for purposes directly connected with our discussions and is also bound to confidentiality as is expressed herein by countersigning this letter or a copy hereof. In such a case XYZ will return to you the countersigned document by recorded delivery post or its equivalent within ten working days of the date of the signature.

XYZ will at no time make, seek to make or allow the making of use of any of the information or items as referred to above which XYZ shall acquire from you in relation to the invention for any purpose other than evaluation of the same without your prior written permission.

The aforementioned obligations of confidentiality shall apply to all information disclosed to XYZ by you except:

1. Information which at the time of the disclosure is in the public domain;

2. Information which after disclosure enters the public domain (other than by the action directly or indirectly of XYZ or of any person in breach of an obligation of confidentiality);

3. Information which XYZ has disclosed as being or in the alternative can prove was in XYZ's possession prior to disclosure hereunder and which was not acquired directly or indirectly from you.

This agreement shall be governed and construed in accordance with the laws of England. Please countersign this letter to confirm your agreement to the terms hereof.

Yours sincerely

Name:

Position:

Countersigned by...

Signed...

Dated..

Countersigned by...

for and on behalf of..

Countersigned...

Dated..

Appendix 6
Companies Offering Service Connections to the Internet

Some companies prefer you to contact them by telephone. We have therefore omitted their addresses.

AOL
Bertelsmann Service Operations Ltd
Gilde House
Eastpoint Park
Fairview
Dublin 3
Tel: 0800 279 1234

BT Internet
Tel: 0800 800001

CityScape Internet services
Alexandria House
9 Covent Garden
Cambridge CB1 2HR
Tel: 01223 566950

CompuLink Information eXchange (CIX)
Tel: 0845 355 5050

CompuServe
1 Redcliff Street
PO Box 676
Bristol BS99 1YN
Tel: 0990 000200

Demon Internet
7th Floor
Gateway House
322 Regents Park Road
Finchley
London N3 2QQ
Tel: 0181 371 1234

Direct Connection
Martin House
1 Tranquil Vale
Blackheath Village
London SE3 0BU
Tel: 0181 297 2200

Easynet
62 Whitfield Street
London W1A 4XA
Tel: 0171 681 4321

Global Internet
Tel: 0181 957 1005

IBM Global Network
Tel: 0800 973000

Microsoft Connection Network
Tel: 0345 000111

PC User Group

PO Box 360
Harrow HA1 4LQ
Tel: 0181 863 1191

PSINet Ltd

Brookmount Court
Kirkwood Road
Cambridge CB4 2QH
Tel: 01223 577167
Mainly for corporate users, but domestic users can connect via their website: www.uk.psi.net

UUnet PIPEX

PO Box 64
Stevenage SG1 2YX
Tel: 0500 474739

Appendix 7
Intellectual Property Organisations and Conventions: Member Countries

The table which follows lists membership of, and signatories to, the following at 3 July 1996:

1. Paris Convention on the Protection of Industrial Property (Paris Convention)
2. Patent Cooperation Treaty (PCT)
3. European Patent Convention (EPC)
4. Eurasian Patent Convention (Eurasian PC)
5. African Intellectual Property Organization (OAPI)
6. African Regional Industrial Property Organization (ARIPO)
7. Berne Convention for the Protection of Literary and Artistic Works
8. Madrid Agreement 1891 (Madrid 1891)
9. Madrid Protocol 1989 (Madrid 1989)

	Paris Convention	PCT	EPC	Eurasian PC	OAPI	ARIPO	Berne	Madrid 1891	Madrid 1989
Albania	✓	✓					✓	✓	
Algeria	✓							✓	
Argentina	✓						✓		
Armenia	✓	✓		✓				✓	
Australia	✓	✓					✓		
Austria	✓	✓	✓				✓	✓	
Azerbaijan	✓	✓		✓				✓	
Bahamas	✓						✓		
Bangladesh	✓								
Barbados	✓	✓					✓		
Belarus	✓	✓		✓				✓	
Belgium	✓	✓	✓				✓	✓	
Benin	✓	✓			✓		✓		
Bolivia	✓						✓		
Botswana						✓			
Bosnia and Herzegovina	✓						✓	✓	
Brazil	✓	✓					✓		
Bulgaria	✓	✓					✓	✓	
Burkina Faso	✓	✓			✓		✓		
Burundi	✓								
Cameroon	✓	✓			✓		✓		
Canada	✓	✓					✓		
Central African Republic	✓	✓			✓		✓		
Chad	✓	✓			✓		✓		
Chile	✓						✓		
China	✓	✓					✓	✓	✓
Colombia							✓		

	Paris Convention	PCT	EPC	Eurasian PC	OAPI	ARIPO	Berne	Madrid 1891	Madrid 1989
Congo	✔	✔			✔		✔		
Costa Rica	✔						✔		
Côte d'Ivoire	✔	✔			✔		✔		
Croatia	✔						✔	✔	
Cuba	✔	✔						✔	✔
Cyprus	✔						✔		
Czech Republic	✔	✔					✔	✔	✔
Denmark	✔	✔	✔				✔		✔
Djibouti					✔				
Dominican Republic	✔								
Ecuador							✔		
Egypt	✔						✔	✔	
El Salvador	✔						✔		
Estonia	✔	✔					✔		
Fiji							✔		
Finland	✔	✔	✔				✔		✔
France	✔	✔	✔				✔	✔	
Gabon	✔	✔			✔		✔		
Gambia	✔					✔	✔		
Georgia	✔	✔					✔		
Germany	✔	✔	✔				✔	✔	✔
Ghana	✔					✔	✔		
Greece	✔	✔	✔				✔		
Guinea	✔	✔			✔		✔		
Guinea–Bissau	✔						✔		
Guyana	✔						✔		
Haiti	✔						✔		
Holy See	✔						✔		

	Paris Convention	PCT	EPC	Eurasian PC	OAPI	ARIPO	Berne	Madrid 1891	Madrid 1989
Honduras	✔						✔		
Hungary	✔	✔					✔	✔	
Iceland	✔	✔					✔		
India							✔		
Indonesia	✔								
Iran	✔								
Iraq	✔								
Ireland	✔	✔	✔				✔		
Israel	✔	✔					✔		
Italy	✔	✔	✔				✔	✔	
Jamaica							✔		
Japan	✔	✔					✔		
Jordan	✔								
Kazakstan	✔	✔		✔				✔	
Kenya	✔	✔				✔	✔		
Korea, Democratic People's Republic of	✔	✔						✔	
Korea, Republic of	✔	✔					✔		
Kyrgyzstan	✔	✔		✔				✔	
Latvia	✔	✔					✔	✔	
Lebanon	✔						✔		
Lesotho	✔					✔	✔		
Liberia	✔	✔					✔	✔	
Libya	✔						✔		
Liechtenstein	✔	✔	✔				✔	✔	
Lithuania	✔	✔					✔		

	Paris Convention	PCT	EPC	Eurasian PC	OAPI	ARIPO	Berne	Madrid 1891	Madrid 1989
Luxembourg	✔	✔	✔				✔	✔	
Macedonia	✔	✔					✔	✔	
Madagascar	✔	✔					✔		
Malawi	✔	✔				✔	✔		
Malaysia	✔						✔		
Mali	✔	✔			✔		✔		
Malta	✔						✔		
Mauritania	✔	✔			✔		✔		
Mauritius	✔						✔		
Mexico	✔	✔					✔		
Moldova, Republic of	✔	✔		✔			✔	✔	
Monaco	✔	✔	✔				✔	✔	
Mongolia	✔	✔						✔	
Morocco	✔						✔	✔	
Namibia							✔		
Netherlands	✔	✔	✔				✔	✔	
New Zealand	✔	✔					✔		
Nicaragua	✔								
Niger	✔	✔			✔		✔		
Nigeria	✔						✔		
Norway	✔	✔					✔		✔
Pakistan							✔		
Panama							✔		
Paraguay	✔						✔		
Peru	✔						✔		
Philippines	✔						✔		
Poland	✔	✔					✔	✔	
Portugal	✔	✔	✔				✔	✔	

	Paris Convention	PCT	EPC	Eurasian PC	OAPI	ARIPO	Berne	Madrid 1891	Madrid 1989
Romania	✔	✔					✔	✔	
Russian Federation	✔	✔		✔			✔	✔	
Rwanda	✔						✔		
Saint Kitts and Nevis	✔						✔		
Saint Lucia	✔						✔		
Saint Vincent and the Grenadines	✔						✔		
San Marino	✔							✔	
Senegal	✔	✔			✔		✔		
Sierra Leone						✔			
Singapore	✔	✔							
Slovakia	✔	✔					✔	✔	
Slovenia	✔	✔					✔	✔	
Somalia						✔			
South Africa	✔						✔		
Spain	✔	✔	✔				✔	✔	✔
Sri Lanka	✔	✔					✔		
Sudan	✔	✔				✔		✔	
Suriname	✔						✔		
Swaziland	✔	✔				✔			
Sweden	✔	✔	✔				✔		✔
Switzerland	✔	✔	✔				✔	✔	
Syria	✔								
Tajikistan	✔	✔		✔				✔	
Tanzania, United Republic of	✔					✔	✔		
Thailand							✔		

	Paris Convention	PCT	EPC	Eurasian PC	OAPI	ARIPO	Berne	Madrid 1891	Madrid 1989
Togo	✔	✔			✔		✔		
Trinidad and Tobago	✔	✔					✔		
Tunisia	✔						✔		
Turkey	✔						✔		
Turkmenistan	✔	✔		✔					
Uganda	✔	✔				✔			
Ukraine	✔	✔					✔	✔	
United Kingdom	✔	✔	✔				✔		✔
United States of America	✔	✔					✔		
Uruguay	✔						✔		
Uzbekistan	✔	✔						✔	
Venezuela	✔						✔		
Viet Nam	✔	✔						✔	
Yugoslavia	✔						✔	✔	
Zaire	✔						✔		
Zambia	✔					✔	✔		
Zimbabwe	✔					✔	✔		

Appendix 8
Select Bibliography

A selection of some relevant titles to read or refer to.

Academic libraries in the UK and the Republic of Ireland, 3rd ed., London: Library Association Publishing, 1994.

All you need to know about UK Internet service providers, Davey Winder, Bath: Future Publishing, 1995 (.net guide 9).

The Aslib directory of information sources in the United Kingdom, ed. by Keith W. Reynard and Jeremy M. E. Reynard, 8th ed., London: Aslib, the Association for Information Management, 1994.
    Has subject index which acts as a guide to organisations active in the field. As well as the details of subject coverage, the entries include information on where to address enquiries and accessibility for non-members, where appropriate.

Be your own boss! How to set up a successful small business, David McMullan, London: Kogan Page, 1994. (Daily Express guides.)

British Venture Capital Association directory, London: BVCA. (Published annually.)

Business plans and financing proposals, revised by Arthur Andersen, London: Business Venture Capital Association, 1993; available free from the Association.

Commonwealth universites yearbook 1996-97, 2 vols., 72nd ed., London: Association of Commonwealth Universities, 1996.
    Published annually; lists address details, staff, research centres

and laboratories, and general information. Includes some international bodies, e.g. Unesco. Has index to subjects covered at each institution.

The complete idiot's pocket reference to the Internet, by Neal Goldman, Indianapolis: Alpha Books, 1994.

Design protection, Dan Johnston, Aldershot: Gower/Design Council, 1995.

Directory of British associations and associations in Ireland, 13th ed, edited by S. P. A. Henderson and A. J. W. Henderson, Beckenham: CBD Research Ltd, 1996.
Has subject index.

Eureka! The book of inventing, Bob Symes and Robin Bootle, London: Headline, 1994.

European research centres: a directory of scientific, industrial, agricultural and biochemical laboratories, 9th ed., Harlow: Longmans, 1993.
Gives full list of company's activities.

Guide to libraries and information units in government departments and other organisations, 32nd ed., Peter Dale (ed.), London: British Library, 1996.
The usefulness of this reference work lies in the editor's inclusion of only those institutions which are "prepared to accept serious enquiries from outside", i.e. most sources are available to non-members, with notes on access details. The emphasis is as indicated in the title: government departments; excludes academic libraries, and commercial libraries except for members of the Patents Information Network and Business Information Network, and European Documentation Centres; includes selected others relevant or of unique importance to the subject, such as official and regulatory bodies and trade associations, e.g. the Aluminium Federation.

Guide to libraries in key UK companies, compiled by Peter Dale, assisted by Paul Wilson, London: British Library, 1993.
> Lists companies by industry sector; has notes on accessibility by members of the general public.

Guide to libraries in London, compiled by Valerie McBurney, assisted by Tony Antoniou and Paul Wilson, London: British Library, 1995.

A guide to venture capital, 3rd ed., London: Business Venture Capital Association, 1995; available free from the Association.

How to set up and run your own business, The Daily Telegraph, London: Kogan Page, 1996.

Internet and comms today, Bournemouth: Paragon. Monthly.

The Internet for dummies, 3rd ed., John R Levine et al., Foster City: IDG Books, 1995.

The Internet for scientists and engineers, 2nd ed, Brian J Thomas, Oxford: OUP, 1996.
> As well as covering access, this work explains how to trace a wide range of sources of information.

Introduction to patents information, 2nd ed, edited by Stephen van Dulken, London: British Library Science Reference and Information Service, 1992.

A Legal guide to innovation, Nabarro Nathanson IP Dept, London: 1994.

Libraries in the UK and the Republic of Ireland, ed. Ann Harrold, 22nd ed., London: Library Association Publishing, 1995.
> Scope: public libraries, academic institutions, and very selected government and special libraries.

Lloyds Bank small business guide, 10th ed., Sara Williams, London: Penguin, 1996.

.net: the Internet magazine, Bath: Future Publishing. Monthly.

NetGuide, Manhattan, NY: Haggety. Monthly.

NetUser, Bournemouth: Paragon. Monthly.

The new Internet navigator: the essential guide to network exploration for the individual dial-up user, Paul Gilster, 3rd ed., New York: Wiley, 1995.

The patent handbook: patent strategies for the small company and private inventor, Noel Stephens, Wotton-under-Edge, Glos.: Perception, 1995.

The practical guide to patents, trademarks, copyright, designs, Laurence Shaw, Birmingham: Bilgrey Samson, c 1996.

The right way to start your own business, Rodney Willett, Tadworth: Elliot Right Way, c 1995.

Setting up a business, Vera Hughes and David Weller, London: Hodder & Stoughton, 1990. (Teach Yourself Books.)

Sources of business angel capital, 3rd ed., British Venture Capital Association, London: BVCA, 1995.

Start a successful business, Rosemary Phipps, London: BBC Books, 1994. (BBC business matters management guides.)

The UK Internet book, Sue Schofield, rev ed, Wokingham: Addison-Wesley, 1995.

The USENET book: finding, using, and servicing newsgroups on the Internet, Bryan Pfaffenberger, Reading, Mass: Addison-Wesley, 1995.

The Venture Capital Report guide to venture capital in the UK and Europe, 8th ed., Oxford: VCR, 1996.

The Which? guide to starting your own business. Commissioned and researched by the Consumers' Association, London: Which?, 1996. (Which? consumer guides.)

The World of learning 1996, 46th ed., London: Europa Publications, 1996.
> Lists address details and staff (in some cases senior staff only) of universities and colleges worlwide. Each country section starts with brief details of learned societies and professional associations. Published annually; no subject index.

Your own business from concept to success, Helen Beare, London: Sheldon Press, 1995.

Index